HPBooks

How to HOTROD Big-Block CHEVYS

by Bill Fisher & Bob Waar

Introduction	2
Let's Get Technical	3
Stock Parts Combinations	8
Metric Equivalent Chart	18
Exhaust	19
Heavy-Duty Clearances	24
Carburetion	25
Ignition	41
Torque Specifications	52
Camshaft	53
Cylinder Heads	77
Pistons	97
Lubrication	111
Turbosupercharging	125
Clutch & Flywheel	126
Bore/Stroke Information	128
Blueprinting & Assembly Tips	129
Variations On A Powerful Theme	148
Heavy-Duty Parts List	150
Tuning Hints	154
Suppliers	159

ANOTHER FACT-FILLED AUTOMOTIVE BOOK FROM HPBooks®

NOTICE: The information contained in this book is true and complete to the best of our knowledge. All recommendations on parts and procedures are made without any guarantees on the part of the authors or HPBooks. Because the quality of parts, materials and methods are beyond our control, author and publisher disclaim all liability incurred in connection with the use of this information.

The cooperation of the Chevrolet Division of General Motors is gratefully acknowledged. However, this publication is a wholly independent production of HPBooks.

Publisher: Rick Bailey; Executive Editor: Randy Summerlin; Editorial Director: Tom Monroe, P.E., S.A.E.; Senior Editor: Ron Sessions, A.S.A.E.; Art Director: Don Burton; Book Design: Max Silten/Silten Graphic, Paul Fitzgerald; Production Coordinator: Cindy J. Coatsworth; Typography: Michelle Carter; Book Manufacture: Anthony B. Narducci; Drawings, Photos: Bob Waar, Howard Fisher, Bill Fisher, Chevrolet Motor Division; Cover photo: Ed Sperko of GM Photographic, Chevrolet Public Relations.

Published by HPBooks
A Division of HPBooks, Inc.
P.O. Box 5367, Tucson, AZ 85703   602/888-2150
ISBN 0-912656-04-2
Library of Congress Card Catalog No. 72-159282
©1971 HPBooks, Inc.   Printed in U.S.A.
20th Printing

Introduction

Since its introduction in 1965, the Mark IV big-block Chevy has appeared in everything from muscle cars, pickup trucks, family station wagons and Corvettes to Can Am cars, full-race boats and a plethora of drag-racing machinery. Shown is 500-CID Pro-Stock drag-race engine. Pair of Holley Dominator 1150s sit atop fabricated aluminum intake manifold. Aluminum heads feature raised exhaust ports for improved exhaust flow. Photo by Tom Monroe.

One of the most formidable large-displacement automotive engines produced in recent memory is the big-block Chevrolet V8. Dubbled "rat motor" and "porcupine," the engines are also known by their cubic-inch displacements: 396, 427 and 454. You should also be aware of the fact that some engines and cars bearing the "396" label are really 402 cubic inches displacement.

In February 1963 Junior Johnson lapped the Daytona Speedway at 166 mph with Chevrolet's Mark II "Mystery Engine." GM dropped out of racing about that time and development of such special engines apparently stopped. As far as the general public was concerned, the only "mystery engines" that they saw for the next two years were a few that had been released to NASCAR racers—which soon went out of action because there were no more parts available to keep them running.

Two years later, a 396-cubic-inch version of the engine was made available in 325, 375 and 425 HP forms. Since original release in February, 1965, the engines have been regularly produced and are readily available in several basic stages of tune from the factory. The design has been proven at the Bonneville salt, the drag strip, the CanAm circuit, in race boats, and in hundreds of thousands of cars and trucks. Never before has this much power been available for such a low dollar cost.

Yet, despite the great number of laurels won by the engine and the increasing number of performance pieces for the large-block Chevy, top engine builders and tuners continue to find ways to squeeze just a few more horses from the engine as the years go by. The modifications needed to produce phenomenal horsepower need not be expensive; they do need to be performed correctly and in many cases must be followed by other changes or modifications. This book is about just that—how to get more horsepower, better performance and longer life from the large-block Chevrolet V8 on the street, at the strip . . . and in other applications demanding more output than comes in the factory package.

Remember the phrase—street and strip—as you peruse this volume. Our suggestions and outlines for increasing performance do not venture into the use of alcohol, super carburetion, injectors, superchargers, magnetos or stroker cranks. This book is about inexpensive, flexible horsepower which can be harnessed for the racetrack or the street.

We are not in the business of manufacturing, selling or proving speed equipment. We have directed this volume at the performance enthusiast who wants to learn how to build and correctly maintain an engine which is suitable for a variety of uses. We have made no attempt to show you how to fit the large-block Chevy into any particular chassis or class of racing.

For the most part, we have confined our discussion to over-the-counter Chevrolet parts— readily available all over the country at reasonable prices. With this information you should be able to build engines which will outperform—both in horsepower and reliability—many which have been thrown together with a full load of "trick" parts. We've done our best to show you how to get maximum performance with minimum bucks.

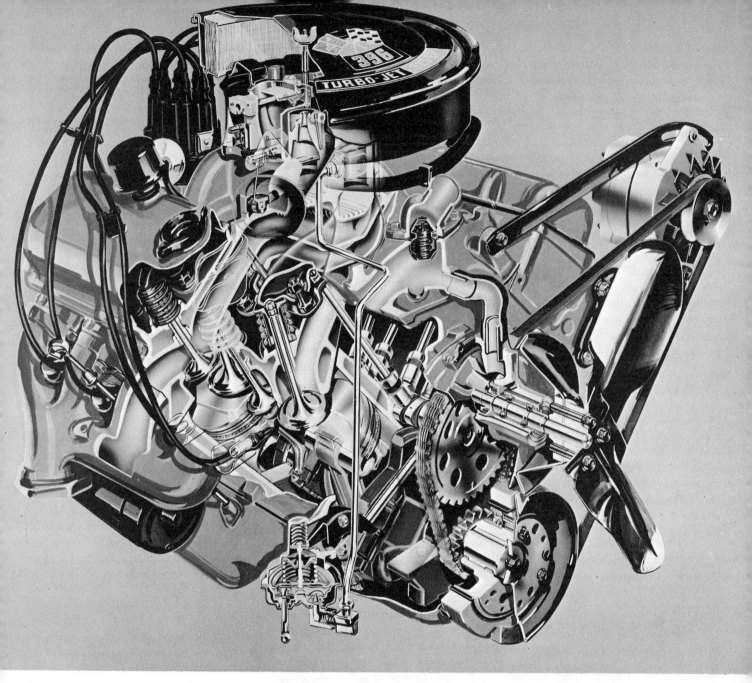

Cutaway view of Turbo-Jet 396-cubic-inch 325 HP engine. This 90-degree cast-iron V-8 has a 4.094-inch bore and a 3.76-inch stroke. This, and the other big-block Chevrolets achieve high volumetric efficiency through a novel cylinder-head and manifold design.

Let's get technical
what's inside the big-block

The 396/427/454 Chevrolet utilizes the most up-to-date design concepts and capitalizes on Chevrolet's past experience in producing highly successful V8 engines. The large-block Chevrolet features high volumetric efficiency and is thoughtfully engineered for high-speed durability.

A complete cast-iron Chevy engine weighs about 680 pounds. Basic engine structures and components are ferrous castings and forgings. Cylinder blocks and heads are cast of high-chromium content iron. The engine is also available in combinations of a cast-iron block and aluminum heads—or an aluminum block with aluminum heads. These weigh 610 and 460 pounds, respectively.

The cylinder block is a short, rigid structure which embodies design features enabling the lower end to withstand unusually high horsepower and torque loads. Cylinder bores are 90 degrees apart and bulkhead thickness above each bearing support is quite massive. Coring provides cylinder walls 0.260-inch thick. There's plenty of "meat" for overboring. Wide-base bearing caps, together with strong bearing bulkheads, provide crankshaft support and clamping. High-performance blocks use four, rather than two hold-down bolts for each bearing cap.

Note: "HI-PERF" cast into a block is not a 100% guarantee that the block has four-bolt mains. A reasonably positive method of identifying a block with four-bolt mains is as follows: check the threaded openings immediately above the oil filter mount. If there are two plugs—one 1/2-inch pipe and the other 3/4-inch pipe, the block probably has four-bolt mains. If it has only one plug, then it probably has two-bolt mains. Of course, the only foolproof way to check for four-bolt main caps is to pull the timing cover or oil pan and start counting the main-cap bolts.

The five-main-bearing crankshaft is forged steel in the high-performance engines, regardless of the displacement. Those engines which are equipped with mechanical-lifter camshafts have forged-steel cranks which have undergone the Tuftride hardening process to attain great fatigue resistance and a tough, hard surface for the main and connecting-rod journals. A few of the big blocks — notably the low-performance 396/402 engines — are equipped with nodular-iron crankshafts. Main bearing loads are reduced through the use of exceptionally large counterweights on all big-block cranks.

Torsional rigidity of the crankshaft is excellent due in part to the 0.60-inch overlap created by the 2.750-inch diameter main journals and the 2.200-inch diameter rod journals. A rubber-mounted harmonic balancer (vibration damper) is bolted to the front of the crankshaft.

Improved main and connecting rod bearings are used with all 396/427/454 crankshafts. The micro-thin babbitt layer, placed over the tough major bearing structure of aluminum alloy, is slightly thicker than previous designs to improve conformability and embedability. This is accomplished while maintaining the high unit load characteristics of the combined bearing surfaces necessary for high-output engines. The four front main bearings and the connecting-rod bearings were originally Moraine 420 — as are the replacements available from dealers — but were changed to the Moraine 400 type because this bearing has better capability to cope with tiny pieces of grit, etc., which are so hard to get out of an engine on a production line. The rear main bearing is the thrust bearing for the crankshaft. It has a babbitt upper which is steel-backed and identified with the letters GMMA. The lower half is Moraine 400 — marked M400. If you take an engine apart and find bearings marked A200 or A220, these are Federal Mogul bearings which are equivalents to the Moraine M400 and M420, respectively.

The upper main-bearing halves are grooved to provide the connecting-rod bearings with a 360-degree supply of oil because the crankshaft main journals — at least on all high-performance cranks — are cross-drilled for this purpose. Thus at any crank position, one of the oil holes is exposed to the groove which is filled with oil under pressure. This assures a film of lubricant under the most severe loading conditions.

The full-pressure oiling system is contained wholly within the cylinder block and head castings. There are no external lines. Oil is supplied by the oil pump, through the full-flow oil filter, to the main oil gallery which extends along the lower left side of the cylinder block.

Oil from the main gallery is routed through diagonal passages in the cylinder block to vertical passages in the bulkheads which connect crankshaft main and camshaft bearings. A feature incorporated in most of the large-block oil pumps hydrostatically balances the pumping gears and relieves the bending moment ordinarily thrust upon the pump drive shaft. Two chambers are formed in the vertical face of the pump housing, adjacent to the drive gear. One chamber is opposite the inlet side of the drive gear, while the other is opposite the outlet side. The pump bottom cover contains a cast passage which connects the oil inlet chamber to the opposite formed chamber.

A second passage, cast in the upper surface of the pump housing, connects the outlet chamber and its opposite chamber. With this arrangement, the high pressure of the outlet side of the pump and the negative pressure of the inlet side are counter-balanced around the drive gear. Elimination of hydraulic forces that normally act perpendicular to the pump drive shaft greatly improves pump durability. In addition, high-pressure lubrication is supplied to the bearing surfaces of the drive shaft by the upper connecting passage.

The large-capacity oil-inlet screen and tube are constructed with minimum restriction, and provide a more direct flow path to the pump-inlet chamber. This eliminates the possibility of oil cavitation, which can cause erratic distributor operation and wear from vibrations transmitted through the pump drive shaft. Smooth pump operation is further enhanced by using smaller gear teeth to reduce the pulse amplitude.

Blocks used for the 375, 425, 430, 435-HP 396 & 427 engines and for 450, 460 and 465-HP 454 incorporate two bosses drilled and tapped above the oil-filter mount for oil-cooler inlet and discharge lines. Check valve 5575416 from a standard six-cylinder truck engine is all that is required to make the system functional — plus the necessary cooler and lines, of course. Oil-pan capacity for the passenger-car engines is four quarts. The Corvette oil pan holds five quarts. A six-quart deep-sump pan 3879633 is available. It is used with a 3879640 tray baffle and a 3964255 pump. The pump has a deep pickup to match the pan depth.

Pistons in the big engine utilize the slipper-skirt principle with webs bridging the two thrust faces and supporting the pin bosses. The pin bosses are moved inboard and thus reduce pin weight while increasing rigidity. Moving the pin bosses inward increases piston-head strength without an appreciable weight increase.

Permanent-mold pistons have balance weights cast into the pin bosses on both sides of the piston pin, instead of below it. This provides clearance for the counterweights on the crank. Balancing pads are also seen on some of the forged (impact-extruded) pistons. The piston head has an inlet-valve relief and a dome shape which accurately corresponds to the combustion-chamber shape.

The combination of slipper-skirt, offset piston pin and auto-thermic expansion control results in a piston with a close bore fit (stock clearance 0.0015-0.002 inch) which runs quietly and has excellent oil and compression control and durability. These pistons are only used in non-high-performance engines. Pistons in the HIGH-PERF engines are of the impact-extrusion type (a type of forging). Thermal expansion control is accomplished by using a barrel contoured piston skirt, which allows eliminating the long slots below the oil-control ring to strengthen the piston head and make the upper skirt more rigid. Cooling is improved because heat flow between head and skirt is uninterrupted for 360 degrees of piston circumference. As in the cast pistons, pin bosses have been moved inboard to strengthen the head and stiffen the pin. Pin bosses are not offset in the forged pistons.

To reduce bore wear and improve compression sealing, barrel-faced top-compression rings are used with flame-blown molybdenum-alloy inlay. Molybdenum seats quickly with the cylinder walls and possesses excellent wear resistance under high-speed and high-temperature operating conditions. Chrome rings stand up best under dirty conditions, while the moly is capable of living with extreme temperatures generated by combustion and friction. A large and effective air cleaner should always be run on the big-block engines to let the moly rings live. Because the stock cleaners are designed with air flow into the carburetor as a primary design parameter, don't be surprised if the car runs quicker and/or faster with the cleaner in place!

For maximum volumetric efficiency current design theory holds that inlet and outlet ports must be as straight as possible, consistent with surrounding component design requirements, and have a minimum change in shape. Inlet and exhaust-port configuration, as well as inlet and exhaust-valve positioning, are worked out with extreme care to produce the optimum induction and exhaust-flow characteristics.

Because of Chevrolet's independent ball-stud rocker-arm arrangement and individual inlet and exhaust porting in the heads, it was possible to cant the inlet valve toward the inlet port. The result is unrestricted inlet ports with fairly uniform cross-section to reduce the changes of direction that the fuel/air mixture must make to enter the combustion chamber. However, because of differences in the angles at which the inlet ports enter the cylinders, two cylinders on each side breathe considerably better than their mates.

Although both the intake and the exhaust valves have been tipped toward their respective ports, the inlet valves have been favored in the carefully thought out design. Intake valves are set at an angle of 26 degrees to the cylinder bore axis and the exhaust valves are set at 17 degrees when viewed from the front or rear of the engine. Viewed from the side of the engine both valves tilt 5 degrees away from this same axis. The combustion chamber is an elongated cavity rotated 18 degrees from the longitudinal axis of the head.

In taking advantage of the tilted exhaust valves, the exhaust ports have a large radius, producing a gradual direction change and an unrestricted uniform cross-section throughout their length.

The merits of this canted-valve cylinder-head design stack up like building blocks. Offsetting exhaust valves toward outer cylinder walls minimizes exhaust port lengths to reduce heat loss into the coolant. Individual porting for each cylinder and the versatility of independent rocker-arm systems provides still another benefit. Keep in mind that inlet and exhaust valves within the same combustion chamber are tilted away from each other along the engine longitudinal axis, as well as the transverse axis. This causes the valve heads to move away from the adjacent cylinder walls as each valve

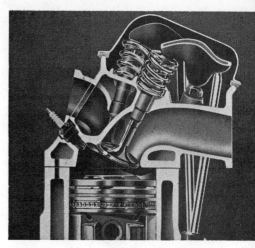

The canted-valve porcupine-head design that makes it all happen. Breathes so well that it's often referred to as a semi-hemi.

opens, instead of paralleling them. Unshrouding the valves in this manner improves incoming and outgoing gas flow which increases the volume of mixture drawn into the cylinder for each engine cycle.

The modified-wedge-type combustion chamber has a large quench area for cooling the mixture so that detonation is minimized and a centrally located spark-plug which results in uniform flame propagation across the top of the piston. The compactness of the chamber improves thermal efficiency by reducing the ratio of chamber surface to chamber volume. In this design there is relatively little piston surface exposed to the high temperatures of combustion; therefore there is less heat to be dissipated through the piston head to the skirts ... and into the cylinder walls.

Combustion chambers of the "open" configuration, as introduced on the ZL-1 and 2nd design L-88 engines — and subsequently in cast-iron form in mid-1971 — are approximately triangular in shape at the gasket surface. The cast-iron version of this head is a performance bonus directly related to GM's march against air pollution. Open-chamber heads have less quench surface, but retained quench where it is needed to ensure good flame travel. Reduced quench area is one of the ways that emissions are decreased. Quench area is that flat portion of the combustion chamber which is separated from the flat part of the piston top at top dead center by only 0.040 inch or so.

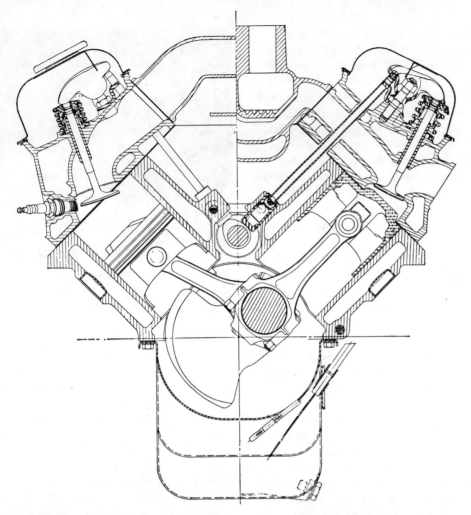

Front section of the 396 shows its rugged, yet extremely simple construction. The water jackets surround the cylinders for nearly their full lengths and provide good cooling for the exhaust-valve areas. These blocks are obviously not "thinwall" castings and can usually be overbored to get displacement increases. This same construction is typical for all of the big-block engines from 366's through 454's. Heads shown in this drawing are of the closed-chamber variety. Intake port shown at left. Exhaust port on right is shown with the exhaust heat riser connection to the intake manifold.

available. These are covered in detail elsewhere in the book — as are gears which were used to eliminate the chain drive on some of the high-performance models.

Much of the explanation behind the efficiency of this entire line of engines is found in the cylinder-head design. Other manufacturers, including Ford, have copied many of the design features . . . as note the design of heads used on the Boss 302 and the 351 small-block Ford engines which have to be called respectable runners in any league.

Bird's eye view of 1965 model 396 shows new Q-Jet carburetor, staggered and splayed valves, and thermostatically controlled fan.

It's a simple engine, so leave it that way

When you look at the design of the Chevy big-block engine, you'll note that there is absolutely nothing there that is not required. The engineers have followed the famous GM engineer "Boss" Kettering's slogan, "Parts left out cost nothing and cause few service problems."

To insure proper cooling, generous passages have been cored around all sparkplugs, valves and valve guides. Valves are so arranged in the combustion chamber so that no two exhaust valves are adjacent. Thus, no exhaust ports are siamesed and the "heat of combustion" is more uniformly distributed throughout the head castings.

Valve guides are pressed into the heads rather than being integral which produces more uniform cooling and reduced guide wear. Rocker studs are of the screw-in variety having about one inch of 7/16 N.C. thread engaging the cylinder head. These studs are installed to match the five degree tilt of both intake and exhaust valves. The rocker arms utilize a ratio of 1.70 to 1 and are beefier, but of the same type introduced by Chevrolet in 1955 on the small-block V8. Hardened-steel pushrod guides near the upper ends of the push rods allow fore and aft travel in relation to rocker-arm motion but permit only minimum side movement.

Connecting rods in these engines are drop-forged steel. They weigh about 30 ounces each. There has been a continual upgrading of the rods since production of the engines started. Differences and methods of identification are clarified later in the book.

The camshaft is chain-driven with a steel crank sprocket and a cam sprocket which is aluminum with nylon covering the teeth. The cam is supported by five steel-backed babbitt bearings. At a diameter of 1.95 inches, the cam journals are exceptionally large. A number of different camshaft grinds are

Several intake manifolds produced by Chevrolet may be used on any of the "rat" engines. And, there's a large selection produced by the various speed equipment manufacturers. These will be discussed in greater detail later, but as an introduction to the component, remarks here will be confined to the two most common Chevrolet manifolds. One is cast iron and the other is aluminum. Save for a weight difference, the major difference is in the size of the runners. The aluminum manifold has larger runners for increased mixture flow in the upper operating range of the engine, thereby increasing top-end horsepower. An exception is the 400 HP 427 3x2 aluminum manifold which has small ports.

Runners of both manifolds are as near equal length as possible, and are configured in smooth, sweeping bends. Uniform cross-sectional areas are maintained throughout runner length. The ratio of wall surface to cross-sectional area is kept to a minimum. Runners for the cast-iron manifold provide optimum torque and horsepower in the intermediate RPM ranges. Simply stated, this means that in the engine midspeed range, runner configuration is such that the induction system pulse initiated by the closing of an inlet valve reaches the inlet valve of the next cylinder to be fired at the approximate time of the valve opening. Thus, the charge of gasoline and air is "helped" into the combustion chamber. The aluminum manifold is similarly designed but sized to provide optimum charging at the high end of the engine speed range.

The plenum floor directly beneath the carburetor is usually ribbed to increase heat transfer from the exhaust crossover passage to the fuel charge. Ribs for both manifolds have been extended into those passages feeding cylinders which tend to run slightly lean. The rib lengths have been precisely determined to direct the mixture so that the fuel charge for all cylinders is as near uniformly balanced as possible. At least, an optimum compromise is reached for the usable RPM range.

To shield the exhaust crossover from engine oil splash and to maintain uniform heat flow, a dead air space is created by casting a second housing over the full length of the exhaust crossover passage for both manifolds.

A permanently sealed fuel pump of simplified design and improved durability is used on all 396/427/454 engines. The large-displacement pump maintains uniform fuel flow to the carburetor; it's not for racing use.

In addition to simple construction, increased displacement improves hot-weather engine operation by rapidly disposing of hot fuel vapors. A single-piece stamped-steel rocker arm is used with a return spring mounted concentric with the diaphragm pull rod. Return-spring operation is in a straight line.

Two different types of carburetors are used for the big Chevrolet engine. Depending on year and horsepower rating, the "porcupine" engine comes equipped with either a Quadrajet or the more conventionally designed Holley four-barrel carburetor. Modifications to both types of carburetors are discussed in some detail elsewhere in the book. Carburetor size or capacity (in cubic feet per minute) is related to engine size and horsepower.

Full-length water jackets around the cylinder bores, generous cooling passages in the cylinder block as well as the heads, short exhaust ports and a high-capacity compact water pump are some of the features of the big V8's cooling system. Efficiency of engine design is such that heat rejection to the cooling system is relatively low.

Basically, a series-flow cooling system is used; the direction of coolant flow is from the front of each cylinder bank to the rear, then upward into the cylinder heads and forward to the thermostat outlet.

The flow pattern is broken only at the spark plugs, where small circular holes pass coolant vertically from the cylinder block into the cylinder heads to insure uniform and continuous cooling of the spark-plug bosses. Large passages direct coolant flow over and around the alternately spaced inlet and exhaust ports, as well as around the exposed exhaust valve guide inserts. Cored passages surround the spark-plug bosses and the full length of each exhaust port with coolant.

The compact and highly efficient water pump has a small-diameter rotor with short runners and large outlets to each cylinder bank. Large passages between cylinder bores and case walls, and full-length water jackets provide uniform cooling to maintain cylinder-block dimensional stability.

The Air Injection Reactor (AIR) system delivers air to the exhaust ports in sufficient quantity to support oxidation of hydrocarbon and carbon-monoxide emissions. This system requires no unusual maintenance and is an integral part of the engine assembly. Major components include an air pump, check valves, anti-backfire valve and special manifolding to distribute the air.

The air pump is a V-belt driven, semiarticulated-vane type, mounted at the front of the engine. Air volume is relative to engine speed. The pump is driven at 1.25 times engine speed. The pump has a displacement of 19.3 cubic inches, and operates in a range of 10 to 50 feet per minute. Carburetor air cleaners provide filtered air to the pump inlet.

From the pump, air is ducted by rubber hose and steel tube to air-distributing manifolds, called combustion-pipe assemblies, which distribute a continuous flow of air into individual exhaust ports. A check valve on each air-distributing manifold limits flow to input air only to protect the remainder of the system from hot exhaust gases if a pump or drive belt fails.

Fuel/air mixtures are controlled during deceleration by a vacuum-actuated anti-backfire valve which adds pump air to the inlet manifold. With the throttle closed, pressure drop causes a diaphragm to unseat the pump air valve and admit air to the inlet manifold. The valve incorporates a pressure-control device to delay closing and prevent rapid oscillation of the valve.

A plate-type pressure-relief valve in the discharge side of the air pump controls flow in the system by venting excess pressure to the atmosphere at about 55 miles per hour. This reduces the volume of air supplied to the exhaust manifold to prevent excessive temperatures from developing in the exhaust system.

AIR-equipped vehicles have radiator fan shrouds and a 195-degree thermostat. Other revisions include higher idle speeds, lean carburetor calibrations, revised initial spark settings, and revised distributor spark advance curves ... all accomplished to help in the air-pollution fight.

Stock parts combinations
similarities make parts swapping easy

396-inch, 425 HP big-block nestles in the engine compartment of a 1965 Corvette.

There are as many ways to "build" a big-block Chevy as there are guys twirling wrenches. Some of the ways are a bunch wrong, and a lot of the methods turn out very good horsepower on the race tracks and on the street. This is what we are after. The following pages are devoted to "building" the big block almost entirely with readily available, reasonably priced Chevrolet parts. We've even included a parts list containing a parts description and part numbers so you'll be armed with more than desire and a blank look when you belly up to the parts counter at the local Chevy emporium.

Keep in mind that following our advice in this section will not guarantee that your engine will be a world beater—far from it. More horsepower can be had with additional carburetion, with more attention to head work than is suggested here and with the substitution of non-Chevrolet camshafts. You should also understand that such "non-stock" variations can easily cause a loss of horsepower just as quickly as they can bend the dyno's torque arm a little further. Build a running "stocker" before gambling with unknown and often unsound combinations.

The 396 Chevrolet

The construction of a high-performance 396 Chevy should start with what Chevy calls the RPO L78 engine. This assembly is distinguished by the four-bolt main-bearing caps and the oil-cooler provision (two threaded plugs just above the oil-filter boss). These plugs can be removed and lines fitted to route oil to and from a remote cooler. If an oil cooler is not used, leave the plugs in place! This L78 engine is equipped with the same large-port cylinder heads and intake manifold that are part of the high-performance 427 package.

Some confusion is bound to result from a slight change made by Chevrolet in 1970 when the L78 was bored out to provide a 402 cubic-inch displacement instead of the previous 396 inches . . . especially since Chevy continues to refer to the engine as a 396. The increase was gained by boring the 4.094-inch cylinders to 4.125-inches. The only other change of any significance is in the intake-manifold design. The 1970 L78 manifold has a lower profile than previous large-block Chevy manifolds . . . which gains hood clearance — but reduces performance throughout the entire engine-RPM range.

425 HP version of the 396-incher uses a single four-barrel Holley carburetor and a mechanical-lifter camshaft. Forged pistons are included among the heavy duty components of the engine. Note the viscous-drive fan which saves HP when the engine is cool.

The 427 Chevrolet

A high-performance 427 Chevy can be put together from any number of several combinations. Let's start from scratch at the Chevy parts store. The easy way out is a RPO L-88 2nd design engine (numbers in the parts list) which comes in the crate with open-chamber aluminum cylinder heads . . . or the $3,000 all-aluminum ZL-1 engine. The 1st design L-88 heads which first saw the light of day late in 1967 were extensively modified when they were released with the ZL-1 and the 1969 and later L-88 engines. Although the intake-port size (3.84 sq.in.) and valve size (2.19 inches) were not changed, the contour of the port was dramatically changed. As the port approaches the valve guide, there is a slight venturi and the "pocket" behind the guide has been filled in so that there is actually less inlet-port section than in the previous heads. This "pocket" apparently caused turbulence which reduced the air flow capabilities. A venturi-like contour at the valve seat smooths the flow through the valve seat opening for an additional flow improvement.

On the exhaust side, the heads have round exhaust ports instead of the previous square ones. The exhaust-valve diameter was increased from 1.84 inches to 1.88 inches. The exhaust-valve seat has a venturi built in.

Combustion chambers are drastically modified from previous versions in these "open-chamber" heads. Accompanying photos show that the second "quench" area by the spark plug has been removed to improve the breathing. The incoming mixture has a much easier path to travel because it does not

All-aluminum 427-CID ZL-1 engine was a Corvette option. It also appeared in some 1969 Camaros.

have to "squeeze past" the valve and the chamber edges to get into the cylinder. There is no more tendency toward detonation—even at the 12:1 c.r. which you typically find in these engines, because quench was retained where needed—farthest from the plug.

The pistons for these heads are especially contoured to work correctly with the open chamber. The TRW forged pistons P/N 3959105 have been beefed up in the area around the pin bosses and skirts. Even so, the weight was kept the same so that the crankshaft counterweighting did not have to be changed.

Save about 150 pounds for only $2,000! That's about all you can expect when you buy a ZL-1 in a crate for about three grand. All of the good features are found in the slightly heavier L-88 — for a lot less money. Unless you are racing in a class where weight is double-important, the cost for one of these would be hard to justify.

Air flow through the open-chamber heads, when combined with the new pistons, is said to be improved by 30% over the first design L-88 aluminum heads or the large-port cast-iron heads. In any event, the combination of camshaft, heads and pistons added a solid 40 HP!

In case you've already traded three broken yo-yo's and a Hudson transmission for a set of these open-chamber heads, you'll need only the L-88 short-block assembly, part number 3970699 (2nd design). It's important to use the correct short-block assembly which matches the heads. The 1st design L-88 short blocks and heads must be used together. Use of the 2nd design heads on a 1st design short block will lower the compression considerably. The 1st design heads absolutely will not fit onto a 2nd design short block unless the combustion chambers are modified to an open-chamber configuration.

Connecting rods for the L-88 are Magnafluxed (but you should never take anyone's word for it—recheck them!), set up for floating pins and are held together at the bottom end by 7/16-inch rod bolts. Several years ago, Chevrolet had some problems in the big-block connecting-rod area. It boiled down to the fact that the top end of the engine was putting out more than the lower end could take. These rods, 3969804 or their predecessor 3959187, have pretty well cured rod-breakage problems in the large-block Chevy. Rods which are used as of 1969 in the L-88's and ZL-1's—also in the high-performance 427 and 454's—will permit occasional use of the engine to 7600 RPM, and extended use at 7000 RPM.

The camshaft on production L-88 engines from 1968-71 is a chain-driven cam which cannot be recommended for street use. These cams are designed for "off-road" applications where open exhausts are the rule. If you are building a street machine, the stock Chevy mechanical-lifter cam for the high-horsepower versions is the one to get if you are shopping at the Chevy dealer's. The last four digits of the part number are usually stamped or cast in the cam by the rear journal. However, L-88 cams purchased from your dealer under part number 3925535 (chain-driven) or 3925533 (gear-driven) may have an 8911 designation. And, these will probably have a groove in the rear cam journal. The non-grooved L-88 cam is not offered as a replacement part. The L-88 cam has a duration of 334 degrees on the intake, with 0.540-inch lift. Exhaust has 364 degrees duration and 0.560-in. lift.

ZL-1 engines are built with cam 3959180 – stamped with 9181 because that's the number Chevy uses for the stick before installing the dowel pin. This cam has 0.560/0.600 lift on the intake/exhaust, respectively. Duration is 347° intake and 359° exhaust. There's even a fuel-injection cam for the ZL-1 and Can Am blocks. It has 0.600-inch lift on both the intake and exhaust with 360°/366° duration. This is P/N 3994094. It is chain-driven, incidentally, as are the cams in all L-88 and ZL-1 engines. Just because Chevy offers gear-driven cams in the heavy-duty parts listing is no reason for you to get excited about installing one — even if you are building an all-out racer!

If you buy a complete L-88 or ZL-1, part of the package is a set of 7/16-inch-diameter pushrods with matching guide plates. If you are putting the engine together from bits and pieces, the inlet pushrods are 3942416 and exhausts are 3942415. Guide plates for the big pushrods are 3879620.

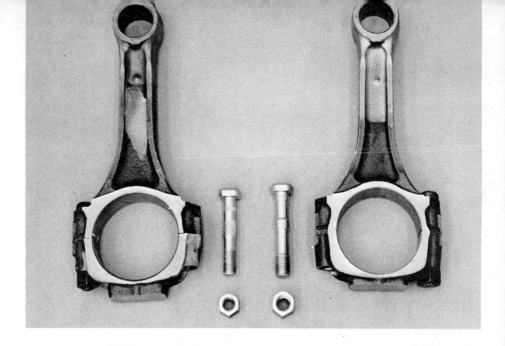

Connecting rod and bolt of 1968 Corvette 427 engine, left, and stronger 1969 rod. Both have ground shank to distribute stretch, are shot peened and Magnafluxed. The 1969 and later rod is also used in the L-88, ZL-1 and LS-7 with the same bolts. LS-6 454's have the good late rod, but with knurled-shank bolts. The knurled-shank bolts are not the ones to use for a high-performance engine.

On some high-performance big blocks, the two holes alongside the front cam bearing are tapped. These are used for attachment of a cam thrust plate when a gear-drive cam is installed. Lifter oiling gallery plugs above cam should be removed when cleaning the block. When replacing the plugs, be sure that you keep all traces of sealant out of the oiling system. Don't use Teflon tape when replacing oiling-system plugs because it is too easy for small pieces of it to get loose and cause damage.

Major components of the 1969 ZL-1 427 aluminum engine. Everything in the engine is designed for GO. Open-plenum manifold kills the low-end torque to a certain degree, but with this much HP and this many inches, a slight reduction in low-end power makes the car more tractable for street operation. Carburetor shown is the big 850 CFM with vacuum-operated secondaries used in automatic-transmission applications. Engine has full-floating pins, 7/16-inch-diameter boron-steel rod bolts and the big-breathing open-chamber aluminum cylinder heads with 1.88-inch exhausts and 2.19-inch intakes.

Front of ZL-1 block. Oil gallery on the aluminum block is adjacent to the camshaft instead of in the block skirt. The two openings just above the cam are the ends of the lifter oil galleries. The cam thrust surface is equipped with two threaded holes (above and below cam) for attachment of cam thrust plate when a gear-drive cam is used. A stainless-steel shim goes between the stock chain-drive cam sprocket and the thrust surface to provide a compatible wearing surface. Otherwise, the aluminum gear riding against the aluminum block would cause disastrous galling.

Parts interchangeability

One of the nice things about working with the big block is the ability to build almost any size engine from any of the original engines by merely swapping stock parts. Although a few points must be watched for, the engines are noted for ease of parts interchangeability and this keeps costs down when you are racing.

Single-snorkle air cleaner and low-profile, staggered-bore iron manifold positively identify the 345/360/390 HP 454 engines. Arrows indicate external balance weights on harmonic balancer and torque-converter plate. 1971 low-HP 454's were equipped with open-chamber-design small-port heads.

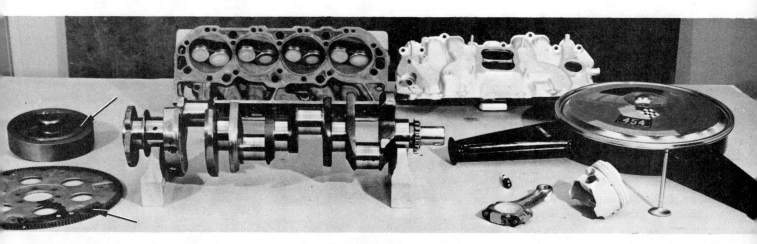

Twin-snorkle air cleaner and aluminum low-rise manifold for Holley four-barrel identify 1970 450/460 HP 454. These engines have 1.88-inch-diameter exhaust valves in large-port closed-chamber heads. All 450/460 HP 454's have forged pistons and rods with 7/16-inch-diameter bolts. 1971 LS-6 options have open-chamber heads.

The 454 Chevrolet

The 454-cubic-inch big-block engine was introduced in 1970. It derives its displacement from a 4.25-inch bore and a 4-inch stroke. Compression ratio in the highest performance open-chambered LS-7 was 12.5:1. Forged pistons with full-floating pins and the top-of-the-line rods with 7/16-inch boron-steel bolts were to have been used in these engines. A lower performance version—the LS-6—had 11:1 c.r. pistons with pressed-in pins in top-quality rods with knurled-shank 7/16-inch bolts. The LS-6 used the closed-chamber cast-iron big-port heads. 1971 LS-6's got the open-chamber aluminum heads. Camshaft in both engines is the same mechanical-lifter stick used in L-78 high-performance 396/402/427 engines since the big block was first introduced. It offers 0.500-inch lift on both intake and exhausts—and duration of 306°/306°. It is a cam which works well with a single four-barrel carb and the kind of exhaust system that's mandatory for the street.

Still lower performance versions were offered with cast-aluminum pistons, hydraulic cam and the Q-Jet carburetor. These were available in 345/360/390 HP versions. The LS-5 RPO describes the 360/390 HP version. Both are apparently the same engine.

1971 454's were only offered in the 365 and 425 HP models. By contrast, the 1970 LS-7 was rated at 460/465 HP and the LS-6 at 450/460 HP. Any way you look at them—and any year you choose—these have to be described as impressive HP ratings for a passenger car.

Careful perusal of the parts list will show that any 396, 402, or 427 with four-bolt mains can be upgraded to LS-6 or LS-7 specifications by merely selecting the correct components and boring the four-bolt 396/402 to 4.251 inches. The 7/16-inch-bolt rods should be used to keep the balance o.k.

The "Super 454" LS-7 was never supplied by Chevrolet. However, the service-parts group stockpiled the 1970 LS-6 block assemblies—both complete engines and partially fitted blocks—and added them to the HD Parts List so that enthusiasts could buy a genuine 11:1 compression ratio engine. As you probably know, the compression dropped to 9:1 on the 1971's, losing 25 HP from the rating applied to the '70. The LS-6 engine for '70 is a 450 HP engine, 3981000. Short blocks bear the designation 3981820 — and a partially fitted block with pistons, pins, rings, and cam and main bearings is 3981806. These engines have a forged and Tuftrided crankshaft, forged pistons with pressed-in pins and a mechanical-lifter camshaft identical to that used in the L-78 427's. If you are dead set on having an LS-7, there's nothing to stop you so long as your cash supply holds out. Change the pistons for the 12.5-er's from the heavy-duty list, modify the rods for full-floating pins if you are so inclined and add the good-guy rod bolts — that's all it takes to be in LS-7 land! However, our recommendation would be to stick with the lower compression, especially in view of the sinking quality of gasoline these days. Stinking quality?

A word of warning is in order for those planning to home-make a 454 incher from a 396 or 427. Minor grinding is required on the bottom of some cylinders to gain connecting-rod clearance. Also—and this is important—think twice before trying to use a 454 crank with a 427 damper and flywheel. Balancing methods are different between the two cranks. Despite what you may have heard, 454 cranks can be internally balanced by adding Mallory metal to the front and rear counterweights. But, this may cost as much as $300. Moldex is one company that can do this—if you're serious. Most balancing firms can do it.

Arrows point to exterior balance weights on converter mount (similar lump appears on flywheel used in manual-transmission models) and harmonic balancer. Forged-steel cranks are the only kind supplied in 1970-71 454-inchers.

Exhaust side of 454 iron cylinder head used on 360/390 and 450 HP models. Small-hex (5/8-inch) tapered-seat plugs were introduced as standard components of all 1970 Chevrolets. Only the aluminum heads continued to use gasketed plugs. Long-reach plug shown for comparison was previously used in iron-head big-block engines.

Notches in the block at the bases of some cylinders may be necessary if a 454 crank is used in a 427 block, or if aluminum rods are used in any of the big blocks.

The 366 and 427 Chevrolet Truck Engines That "366" number may sound strange to many readers, but it is the displacement of a bonafide big block which is used in many of Chevy's larger trucks. It combines a 3.935-inch bore with the 3.76-inch stroke that you are accustomed to in the 396/402 engines. Blocks used for this 366 and for the 427 truck engines are 0.4 inch taller to accommodate the extra piston ring which appears on truck pistons. The blocks measure 10.2 inches from deck to crank centerline, as compared to 9.8 inches for passenger-car engines. Another area of difference is the manifold-seal area or rail at the front and rear of the block. This is considerably longer on the truck blocks. That should give you the clue that truck manifolds are wider because the 0.4" added height moves each head upward and outward on the V.

All big-block parts except pushrods, pistons, manifolds and distributor housings will fit into these engines. And, a car-type manifold could be used if you made thick spacers for each side of the manifold.

If you found one of these engines at a good price, it could be a good buy, so long as you keep in mind the fact that special pistons would be needed for high-performance applications. You should expect to bore any 366/427 truck engine that you find because trucks typically rack up high mileages in a big hurry. We'd also suggest an extra dose of care in checking such blocks for main-saddle and cap cracks — and for cylinder-wall damage.

Most of the 366/427 truck blocks are marked "TRUCK," but you cannot count on this for a sure identification, so check any big-block carefully before buying it. Some of the truck engines had gear-driven cams, but the chain-drive cams will directly interchange. These blocks also are equipped with four-bolt main-bearing caps.

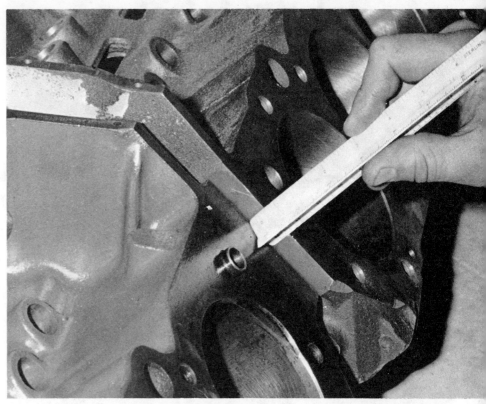

Big-block 366 and 427 Chevy truck engines have blocks with 0.4-inch taller decks to accommodate the extra height required for the four-ring pistons used in trucks.

This chrome-plated example of big-block power was found in a drag boat. Injected or carbureted, these are the most popular large-displacement performance engines ever to be used in drag or ski hulls. Chevys "rule" on water as on land.

Reynolds high-silicon 390 alloy—as used in the Vega engines—was used to cast the linerless block at the top. Engine is compared with Chevrolet's Can Am block at bottom. Chevy block has iron liners. ZL-1 open-chamber-type heads are used on both engines. Note extra hold-down studs in the tappet chamber. Photo courtesy Reynolds Metals.

The 482 Chevrolet Turbo-Marine

You read a lot about these engines, which is unfortunate because your appetite was probably whetted for something that's just not available. Apparently 25 of these engines were made for Kiekhaefer and destined for off-shore long-distance racing. A few cranks, rods and pistons were available for a short while, but we could not find any of these parts in a rather thorough search of logical suppliers around the country. Don't run up your phone bill looking for the pieces. We've already done that for you.

A 454 crank or a 396/402/427 crank stroked to 4.25 inch will provide the 482-inch displacement with the stock 427/454 bore.

Tests on one of these engines equipped with two turbosuperchargers appear on page 125.

Reynolds Aluminum Blocks

In addition to the GM-supplied aluminum blocks with cast-iron sleeves, Reynolds Aluminum of Richmond, Virginia—suppliers of the 390 alloy used in the Vega 2300 engine block has made some sleeveless big blocks for use in the McLaren cars in 1967-71. This team dominated Can Am racing throughout 1969-70—using the Chevy-made Can Am blocks with 4-7/16" bore. For winning the October 18, 1970 Laguna Seca Can Am race, Denny Hulme used the Reynolds all-aluminum block without iron cylinder liners.

According to Reynolds' Director of Automotive Development Harold Macklin, this design could allow building big-block Chevys with more than 500 cubic inches displacement. Because there are no iron liners, the blocks are 15 pounds lighter than the GM Can Am blocks. Perhaps more important than the weight saving is the fact that the faster heat dissipation without the sleeves—which act as heat "barriers" because of the dissimilar metals and non-molecular contact between the sleeve and block—allows smaller water jackets because they don't have so much work to do. Thus, the bores can be made larger.

Reynolds block bores start at 4-7/16 inches—and larger bores are possible because there are no iron sleeves. Pistons, like the Vega's, are electrolytically plated with cast-iron and run directly on the aluminum bores. Blocks are made from Reynolds' 390 alloy. 1971 McLarens are using a 4-1/2" bore with 3-3/4" stroke to get 477 cubic inches displacement—as opposed to the 465 inches available with the same stroke in the Chevy Can Am block. The 454 crank could be used in one of the 4-1/2" bore Reynolds blocks to give 509 CID—or 524 CID with a 4-9/16" bore. At the beginning of 1972, Chevrolet started marketing a 390 aluminum alloy block. The cost is just about what a linered block cost in 1971. Most of the Can Am racers are switching to the lighter block which is sold with a bore of 4.44-inch. Pistons are available in 0.010, 0.060, and 0.070-inch oversize to accommodate a 4.5-inch bore and 4.5-inch plus 0.10-inch cleanup. Linered blocks are no longer being made because the durability of the 390 alloy component is now superior

Colin Beanland with one of the Reynolds all-aluminum-block McLaren racing engines. We made this Reynolds-supplied photo a big one because it contains a wealth of detail which the serious big-block enthusiast will appreciate. Pay no attention to the injector-stack arrangement because they were just jammed on for the photo taking. Note the big 2-1/8" header pipes leading to the long collector. Dry-sump oiling system clearly shows one of the pickups at the rear of the cast-magnesium pan. The four-stage Weaver Brothers dry-sump pump with the toothed-belt drive is also clearly illustrated. No iron liners are used in these blocks. Iron-plated forged-aluminum pistons run directly on the Reynolds 390 alloy bores.

Metric Chart

METRIC CUSTOMARY-UNIT EQUIVALENTS

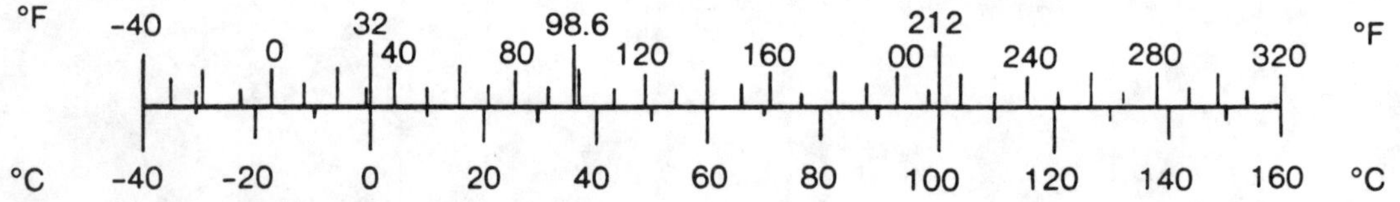

Multiply:	by:	to get:	Multiply:	by:	to get:
LINEAR					
inches	X 25.4 =	millimeters (mm)	X	0.03937 =	inches
miles	X 1.6093 =	kilometers (km)	X	0.6214 =	miles
inches	X 2.54 =	centimeters (cm)	X	0.3937 =	inches
AREA					
inches^2	X 645.16 =	millimeters^2 (mm^2)	X	0.00155 =	inches^2
inches^2	X 6.452 =	centimeters^2 (cm^2)	X	0.155 =	inches^2
VOLUME					
quarts	X 0.94635 =	liters (l)	X	1.0567 =	quarts
fluid oz	X 29.57 =	milliliters (ml)	X	0.03381 =	fluid oz
MASS					
pounds (av)	X 0.4536 =	kilograms (kg)	X	2.2046 =	pounds (av)
tons (2000 lb)	X 907.18 =	kilograms (kg)	X	0.001102 =	tons (2000 lb)
tons (2000 lb)	X 0.90718 =	metric tons (t)	X	1.1023 =	tons (2000 lb)
FORCE					
pounds—f(av)	X 4.448 =	newtons (N)	X	0.2248 =	pounds—f(av)
kilograms—f	X 9.807 =	newtons (N)	X	0.10197 =	kilograms—f

TEMPERATURE

Degrees Celsius (C) = 0.556 (F - 32)   Degree Fahrenheit (F) = (1.8C) + 32

°F -40   0   32 40   80 98.6 120 160 212 00 240 280 320 °F
°C -40 -20   0   20   40   60   80 100   120 140 160 °C

ENERGY OR WORK

| foot-pounds | X 1.3558 = | joules (J) | X | 0.7376 = | foot-pounds |

FUEL ECONOMY & FUEL CONSUMPTION

| miles/gal | X 0.42514 = | kilometers/liter (km/l) | X | 2.3522 = | miles/gal |

Note:
235.2/(mi/gal) = liters/100km
235.2/(liters/100km) = mi/gal

PRESSURE OR STRESS

inches Hg (60F)	X 3.377 =	kilopascals (kPa)	X	0.2961 =	inches Hg
pounds/sq in.	X 6.895 =	kilopascals (kPa)	X	0.145 =	pounds/sq in
pounds/sq ft	X 47.88 =	pascals (Pa)	X	0.02088 =	pounds/sq ft

POWER

| horsepower | X 0.746 = | kilowatts (kW) | X | 1.34 = | horsepower |

TORQUE

pound-inches	X 0.11298 =	newton-meters (N-m)	X	8.851 =	pound-inches
pound-feet	X 1.3558 =	newton-meters (N-m)	X	0.7376 =	pound-feet
pound-inches	X 0.0115 =	kilogram-meters (Kg-M)	X	87 =	pound-inches
pound-feet	X 0.138 =	kilogram-meters (Kg-M)	X	7.25 =	pound-feet

VELOCITY

| miles/hour | X 1.6093 = | kilometers/hour (km/h) | X | 0.6214 = | miles/hour |

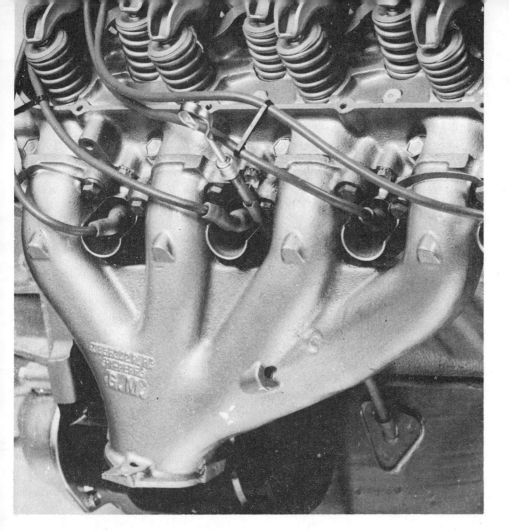

Exhaust
getting rid of waste ups HP

Corvettes have the best exhausts offered on rat motors. And, unless you are adding a severe cam and running open exhaust, you should leave them on your engine to avoid the noise and expense of tube-type headers. We'll tell you how to get good performance with the manifolds and stock-type mufflers. The quiet system will let you use your car's performance.

THE STOCK SYSTEM

Big-block cylinders exhaust through short exhaust ports into a cast-iron exhaust manifold which is connected to the exhaust system. A thermostatically operated exhaust heat-riser valve directs exhaust gases through the intake manifold until the engine warms up. If you block off the heat with blocker-type intake-manifold gaskets—or if you have blocked off the heat by installing steel shims between the head and gasket at the heat-riser port—you may want to block open the heat-riser valve or replace it with spacer block 3796797.

If you want to keep your car *very quiet*, then the power-robbing stock exhaust manifolds are the answer. Their cast-iron construction effectively dampens exhaust noise, whereas tubular headers are almost "bell-like." If you are committed to the use of iron manifolds and the HP loss which they inevitably cause, you are probably "stuck" with the manifolds that came on your car because the best manifolds—supplied on Corvettes—won't clear the chassis of the other Chevys.

Now let's get our heads on straight and look at the real problem. We have to drive our cars on the street, right? And tube-type headers are so noisy that they'll give you a headache with their constant ringing. You didn't know that? Well, it's the gospel truth and you'd better believe it! Headers give 25 to 50 more HP *with open exhaust*. If you have to use mufflers and a cam no stouter than the street-mechanical, skip to page 22 where we tell you how to get what you really want: high performance for low bucks. All of the *absolute musts* that we talk about for headers relate to *competition with open exhaust*. End of lecture.

TUBULAR HEADERS

If the big-block has a single most-obvious Achilles Heel, it is the exhaust manifolding. Even the high-flow manifolds used on high-HP and Corvette engines are restrictive as compared to the tubing-type headers which can be purchased for reasonable prices. In fact, the addition of a set of exhaust headers should be the very first addition that any big-blocker even considers. It is HP that is the very cheapest that you can buy—lots of horses for few bucks.

As an average, you can figure that headers will add 25 to 50 HP to your big block—more if the engine is really a hot one. Your big-block Chevy can never reach its full potential if you insist on running its exhaust through the stock cast-iron manifolds—even the free-flowing ones that are on Corvettes.

One test run by Ronnie Kaplan on the 1970 LS-6 450 HP (factory rating) 454 engine whistled up a mere 380 HP with the stock manifolds, yet the addition of tube-type headers tacked on a 74 HP increase to give 451 HP. This could give you more incentive to buy headers with your first bolt-on dollars.

Tubular headers are an absolute must for the big-block Chevy. They scavenge the cylinders of most of the burnt gases which are being expelled from the engine so that the engine draws in a completely fresh charge of fuel and air that can be ignited to make maximum power. With the restrictive stock exhaust

Those three 2's will never deliver factory HP ratings with these restrictive iron manifolds in place. Tubular headers are the path to more power at low cost.

manifolds, the burnt gases are not only not completely drawn out of the cylinder —part of the gases may be bounced right back into the cylinder before the exhaust valve closes, thereby diluting the incoming charge and reducing the amount of power which can be produced. Part of this problem is created by the fact that the exhaust manifold is a "log" with all of the ports on a head dumping into it. The log size and shape prevent isolating the ports from one another, so there is every good chance that the manifold will be running under pressure which is sufficient to drive part of the exhaust back into the cylinders when pressure in the still-open cylinder is less than that in the exhaust manifold.

On 1973-and-later cars, exhaust-gas recirculation is used. A portion of the exhaust is recycled through the engine to reduce emissions. This really *destroys* engine performance. Such solutions may make pre-1973 cars the hot setup for years to come. We predict that these cars will become performance classics—at least until Detroit engineers figure out how to clean up internal-combustion emissions without making the engine run worse.

You should also note that headers usually cause the engine to run cooler. The stock exhaust manifolds are large heat sinks which store enormous amounts of heat right alongside of the cylinder head. This places an extra burden on the cooling system and also raises the temperature of the underhood air—detracting from performance unless the carburetor is being fed cool air from the cowl or front air scoops. The engine also runs cooler because the carburetor is jetted richer to take full advantage of the extra HP produced with the headers. A portion of the mixture is often carried through the combustion chamber and out with the exhaust on the overlap, cooling the exhaust valve, guide and port in the process.

AVAILABLE HEADER SYSTEMS

Because a set of headers is one of the best ways to add HP to your big-block at reasonably low cost, you'll want to buy the best that you can afford. For a street machine, you can use almost any of the systems with 2-inch pipes leading from the exhaust port flanges to the collector.

Got a problem with headers not snugging up to the head or loosening up after running? Try cutting the flange in a couple of places with a hacksaw to allow more flexibility. Looks backwards to you? You are right — this engine is installed on a dynamometer. Such testing nearly always has the headers pointing to the front of the engine. Note plumbing for remote filter and large line for direct-reading oil-pressure gage. Deep sump indicates that this is a drag engine.

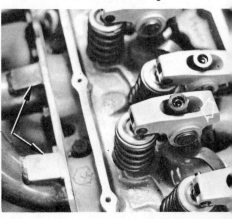

Roller rockers on the big-block Chevy are often used for high-RPM engines which see race-track service but they should not be used on the street due to inadequate oiling. The small hole (outline arrow) feeds oil to the rocker from the pushrod. Other arrows indicate where a header pipe can be braced to the flange with a light piece of plate. Header cracking is a problem in boat and off-road racing.

Most headers are sold through speed shops and mail-order houses at some kind of a discount off of the list price. This being the case, you'll do well to shop when you are buying. Headers are not something that you can buy through the Chevy dealer at a decent price—unless he is a high-performance dealer who is stocking some brand other than the ones that Chevrolet buys from Kustom Headers and catalogs with Chevy part numbers for the L-88/ZL-1 Corvettes.

The factory engineers' recommendation is as follows: "A satisfactory tuned open exhaust is mandatory to extract maximum torque and horsepower from these engines. Correct dimensions for such a system are 2" outside diameter by 36"-long head pipes collected in a group into a 3-1/2" to 4" collector. The 3-1/2" size can be used only if the collector is not more than 24" long. Otherwise, the collector should be 4" diameter. Four-inch tailpipe is preferred for any installations requiring more than 36" of tailpipe. Several header systems designed to these dimensions are currently being marketed by speed shops and high-performance parts manufacturers. Fuel-injected engines respond favorably to 2-1/8" OD head pipes 36" to 40" long." If you are buying a system—or fabricating one from a kit—you'll want to stick close to the recommended dimensions.

One of the big factors in your decision of which ready-made header to buy —in addition to price—should be how easily the headers can be installed. How well do they fit? Here, you'll want to check with your friends and talk to owners of similar cars at the drag strip. Ask the people you meet how much trouble the header installation was—if they did it themselves. You might find out that the reason the local speed merchant offers such a great deal on his remaining header stock is that he's stuck with 'em because the darned things take so much labor to install that you'll spend as much with the local muffler shop getting them installed as the headers are costing in the first place—maybe more. If the headers are anything other than a simple bolt-on job, then the installation cost has to be added in when you are making price comparisons. For that reason, the higher-priced set of headers may well end up costing you less in the long run.

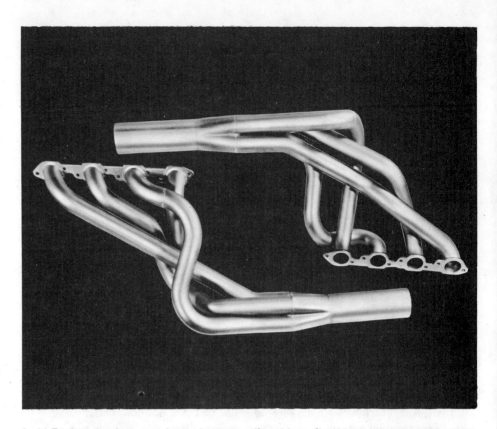

Stahl Engineering lets you choose between adjustable or fixed-length primary pipes with diameters of 1-7/8-inch, 2-inch or 2-1/8-inch. Collectors are available with diameters from 3 inches to 4 inches in 1/4-inch increments. Units shown here are for 1965-71 Chev passenger cars and 1967-69 Camaro.

The factory recommendation in the preceding column is designed to give a very broad torque range. Big-block drag racers have tended to ignore the need for a broad torque range and thus have tended to stick with flat-tappet cams and big primary header tubes. They have been reaching only for peak HP without regard for the acceleration improvements which may be available with a broader torque curve. So, the following recommendations from Jere Stahl should be considered as relating to the search for peak HP for drag purposes. Engines built for a broader torque range — as with roller-tappet cams — might require different combinations.

Stahl and the other experts agree that the 396 CID small-port engine likes 1-7/8-inch primary pipes with a 3-1/4-inch collector as long as the cam is stock or mild. When cam timing gets out of the mild category, a 2-inch primary system with the same collector size is the choice.. 427's (open-chamber) run well with 2-1/8 primaries and 3-1/2 collectors. When cam timing is increased — as with Chevy 3959180 — primary pipes can be shortened. Primaries should usually be not less than 28 — or more than 38 inches long. Collector length may vary from 13 – 17 inches. Engines larger than 427 CID use the 2-1/8 primaries and 4-inch collectors. Regardless of the application, this all gets pretty tricky. Stahl, Hooker and others offer header systems configured so primary and collector lengths can be varied. All this is pure race-car stuff and doesn't really relate to the street.

Cast-iron manifolds such as these are the most restrictive ones offered for the big-block. However, with a bit of effort on your part, you can have good performance for the street by combining these manifolds with the correct exhaust pipes and mufflers. And, you'll have a quiet-running engine that does not draw undue notice every time that you bury your foot in the carburetor. If you are tempted to buy headers for a street-only machine, make sure that you listen to a similar engine which is so equipped. Headers are noisy and you might not like them!

No one header manufacturer makes an ideal header for every chassis and engine combination—nor will he claim to do so. There are two types of headers which are usually offered: under the chassis and through the fender well. The fender-well types are in evidence at the drag strip because they apparently will make more HP. However, for a car to be run on the street, the tuck-under-the-chassis kind are preferred. If you live in a state which has roadside inspections, then you will probably have to get headers with the A.I.R. exhaust-air-injection provisions to match those which were on the original exhaust manifold. Californians take note!

Although the header manufacturer may offer some super-slick glass-packed straight-through mufflers which are guaranteed to sound like a 747 under full take-off power—avoid these like the plague. The kind of muffler that you can use on the street is fully described in the following section.

Decals should be considered as meaningless indications that hero racers are being paid $$$ to stick them on their cars. This is especially true in the case of headers. Before you rush to buy the type of headers claimed by your drag hero's decal, check his car very carefully to see if his headers are really the kind that *you* can buy. They could be made by another firm whose decal is not displayed. Don't be misled.

MUFFLERS

You'll want two—one for each side of the car. Not for the extra noise that this makes—but to reduce the restriction in the exhaust system. It is absolutely essential that you replace original skinny exhaust pipes. At least 2-1/2" exhaust pipes should be used from the header collector to the muffler. Get that muffler as far back under the car as you can—and then make the tailpipe at least 2" diameter.

Which mufflers should you buy? As already stated—not the straight-through glass packs unless you have a bottomless wallet and can put up with all of the harassment that this will bring from the men in blue. Keep it quiet!

You'll have more fun with your car's performance as a result. Some possibilities you may want to consider are the Corvair Spyder mufflers, P/N 3869877. These are possibly the least-restrictive core-type mufflers you can buy. They do a fairly effective job of silencing, especially if you add a balance pipe as described later in this section. Other possibilities to consider are the 1963-64 Chevy 409 high-performance mufflers, P/N's 3852809 and 3852810 for left and right, respectively. You might also want to consider high-performance Pontiac mufflers from their single-exhaust models and the low-restriction mufflers that Chrysler uses for street hemis.

The correct selection of a pair of mufflers can make a big difference in performance if they are connected through 2-1/2" head pipes as already described. Some performance difference can even be obtained when the free-flowing parts are tacked on behind the stock cast-iron manifolds. Tests made by Dick Griffin of Lansing, Michigan, jumped a 396 Camaro 350 HP with Turbo-Hydro from 14.8 seconds ET at 97 mph to 14.0 seconds ET at 102 mph—with the stock cast-iron manifolds. Such improvements are certainly worthy of your consideration.

Making it quieter

If you get the muffler system installed behind your headers and find that the whole affair is just too cotton-pickin' loud, run a balance tube ahead of the two mufflers between the two exhaust pipes. The larger you make the balance tube—the quieter the system will become. Use of a balance pipe to connect the two sides dampens the low-frequency pulses to a lesser sound-pressure level. The factory puts the single exhaust system on their bread-and-butter cars. It "speaks" at 16,000 pulses a minute when the engine is turning 4,000 RPM, or 266 pulses per second. This sounds smoother to our ears than the same exhaust split into two four-cylinder sets—each talking at a 133-pulse-per-second rate. Most dual arrangements are cop-calling stereo systems. But the balance pipe ahead of the mufflers allows each side to speak more softly—at the 266-pulse-per-second rate—and back pressure is reduced still further and the noise level approaches that of a stock single-muffler system.

Don't regard noise lightly if you want to use your big-block engine's full performance potentialities. A quiet exhaust system attracts much less attention than a noisy one. The same is true of carburetor-intake noise because it can be louder than the muffled exhaust under certain circumstances. Don't neglect this point because the quieter the exhaust—the less traffic tickets—of all types. We have proved that time and again on street-driven bikes and cars.

EXHAUST CHANGE REQUIREMENTS

Any time you change the engine so it breathes more freely, carburetion and ignition timing changes may be needed to get maximum performance from the new combination. When headers alone are added to a factory-assembled engine, some intake-mixture richening may be required IF the carburetor is too small. It all depends on the pulsing through the carburetor which is caused by the exhaust change. Read the plugs frequently as a guide. Slight retarding of the ignition may be required.

EXHAUST SYSTEM APPEARANCE

Keeping your exhaust system "looking young" is no difficult chore if you will clean off the paint that the header maker puts on—usually by sandblasting—and apply Sperex VHT enamel to the pipes, flanges and collectors. The parts will look even better if you grind off any ugly welds before painting. VHT is the paint used on all sorts of race cars and boats. It is available in black, white and a wide array of colors.

TUNED EXHAUST SYSTEMS

We won't go into the construction of tuned exhaust systems because that's the subject of another book. You might want to read it someday. Get a copy of Philip H. Smith's "THE SCIENTIFIC DESIGN OF EXHAUST AND INTAKE SYSTEMS." In passing, we'll note that the four-into-one collector systems being made for the big-block by most header makers are pretty close to what would be optimum. Note Chevrolet's recommendation earlier in this chapter.

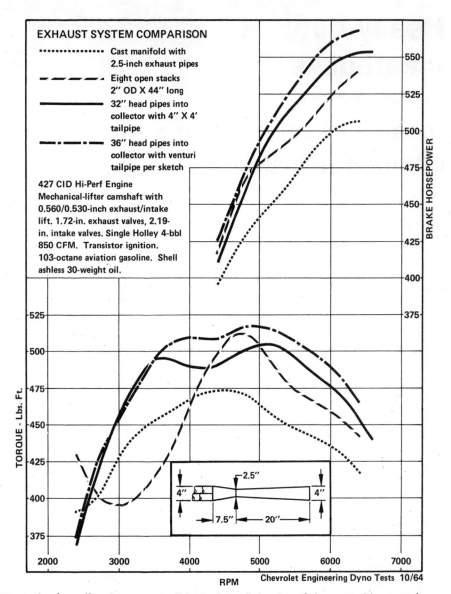

"Venturi exhaust" collector is a valid principle IF the size of the venturi is opened up to accommodate a freer-breathing intake system such as dual quads or fuel injection.

Tuned-length pipes to the collector and various collector lengths and diameters can be used to fine-tune an exhaust system. If you are drag racing, then such adjustable systems could be worth the effort. We've noted that Grand Prix racing teams—such as Ferrari—don't hesitate to use such tactics in fine-tuning their cars for particular courses. Stahl Engineering pioneered the use of adjustable headers for production cars here in the U.S.

The accompanying graphs from Chevy show what you give away when you run straight stacks on your big block. Do you really have that much more HP than your competitor? You'll probably have to make that reverse-cone collector and megaphone affair yourself because none of the header makers offer them as of 1971.

You might think that tuned-length straight stacks would be the best-possible answer to the problem. That has not turned out to be the case. In fact, the straight stacks are very "peaky"—providing a large increase in power at a particular RPM—and detracting from power output throughout the rest of the engine range. The trend is definitely away from straight stacks for racing cars. You certainly won't see any on Can Am cars. A lot of rail-type dragsters are now equipped with collector-type systems. We predict that even the funny cars will someday switch from their straight stacks.

23

Heavy-duty clearances

build according to these and your engine will live

RECOMMENDED CLEARANCES

Piston-to-Bore:	0.0065-0.0075" measured at *centerline* of wrist pin hole, perpendicular to pin. Finish bores with No. 500 grit stones or equivalent (smooth).

Piston-Ring Gap:	Minimum end clearances: Top 0.022
	    2nd 0.016
	    Oil 0.016

Wrist Pin:	0.0004-0.0008" in piston, 0.0005-0.0007" in rod. End play 0-0.005" preferred.
Rod Bearing:	0.002-0.003", side clearance 0.015-0.025" minimum preferred per pair of rods.
Main Bearing:	0.002-0.003", minimum preferred, 0.005-0.007" end play.
Piston to Top of Block: (Deck Height)	0.0-0.005" average above deck, with piston centered in bore. Deck height specified is for a 0.040" (compressed) Victor composition head gasket. If thinner head gasket is used, deck height may be increased accordingly. For best results, piston-deck-to-cylinder-head clearance should be established at 0.035-0.040" with piston centered in bore.
Valve Lash:	0.022 intake, 0.024 exhaust-*hot* for L-88 and ZL-1 type camshaft with aluminum heads. 0.024/0.026 for cast-iron heads. Cold lash on all-aluminum engine is 0.012 intake, 0.014 exhaust.
Valve-to-Piston Clearance:	0.020" exhaust, 0.015" intake at 0 valve lash. NOTE: These are absolute minimum clearances for an engine to run below the valve-train-limiting speed of 7600 RPM. If you intend to run up to valve-train-limiting speed, more clearance is essential as described in the text.

Jerry Thompson at right is shown at the 1969 A.R.R.C. Daytona meet in a big-block-powered Corvette coupe. In the '69-70 seasons, he and partner Tony DeLorenzo grabbed off 23 firsts in 23 starts in an 18-month period. These two racers offer special chassis parts for Corvettes, Camaros and Vegas through their company, Troy Promotions. Their parts list includes the solid bushings needed for autocross suspension mods, competition swaybar kits and track bars.

Carburetion
a little is better than a lot

1969 was the last year that you could buy the three-carb-equipped (3 x 2) Corvette big-block engines. Both the 400 and 435 HP 427-cubic-inch engines had this arrangement in '69. By 1970, the factory supplied all high-performance engines with a single Holley four barrel. If someone tries to sell you a 3 x 2 — or trade you for your four-barrel high-rise manifold — pass! There's no reason whatsoever to bolt this kind of complexity onto your engine — not when one four barrel will "make" as much HP. If you are building a drag car, the 3 x 2 has long been obsoleted by the dual tunnel-ram and injectors.

Most of the big-block Chevys are equipped with a Rochester Quadra-Jet (Q-Jet) or a Holley carburetor. Both four-barrel carburetors have two smaller primary "barrels." A very few low-HP engines have a two-barrel Rochester carburetor. We have assumed that you are not interested in these two-barrel carburetors and carburetion info in this book is devoted almost entirely to the four-barrel carburetors. Although we have shown photos of some of the hot machinery with dual four-barrel carburetors for drag racing and injectors for Can Am and boat racing, our emphasis is on streetable equipment. Our discussion of carburetion is primarily limited to a single four-barrel installed on either a factory manifold or one of the unusually efficient Edelbrock Tarantula manifolds or Holley Street Dominator or Strip Dominator manifolds.

The carburetor is mounted in the center of the engine on an aluminum or cast-iron low-riser manifold, or on an aluminum high-riser manifold. Holley four barrels are always mounted on an aluminum manifold. These were high-riser types until 1970 when low hoods forced the use of low-risers in all models. High-risers won't fit under any 1970-72 hoods except Chevelles—unless the hood is "bumped out." Tall hoods and high-rise manifolds are listed as service packages. The manifolds are usually easy to get, but don't hold your breath while waiting for a high-rise hood to show up after you've ordered one. You could be in for a long, long wait.

Because carburetion systems on the big-block Chevys are inextricably tied into the emission controls, it is important to have a service manual for your particular car so that you can learn

Rochester Q-Jet is equipped with tiny primary venturis and hugh secondaries. Secondary operation is controlled by an air valve to give good cruising gas mileage with adequate full-throttle power. While it is the best performance/economy carb in the industry and is used on millions of GM cars, it is a bit too small for true high performance from the big-block Chevy. Tuning details are provided later in the chapter.

to understand the workings of the various emission-control devices. Some of these include transmission-controlled vacuum spark advance (TCS), specialized mechanical-advance curves in the distributor, carburetor-air heat, and on some models, a device which cuts in the air-conditioning compressor as you turn the ignition switch off to load the engine so it will shut off if dieseling (running-on) occurs.

Lest you think that all of these items have been added to make your life harder, think seriously and unselfishly about the clean-air problem which must be solved if we want to continue living on Planet Earth. Make yourself part of the solution—instead of part of the problem—by getting acquainted with the systems and what each does.

We won't try to kid you —almost anything which makes an engine "cleaner"—so that it produces less emissions—makes the engine run rougher and produce less HP. Short-circuiting or deactivating these control devices can be helpful for winning at the drag races, but in California and certain other states, roadside inspections are made without warning to check that all systems are operable. The fines for deactivating the systems for *road use* are severe and getting more severe every year. Don't put yourself in a financial bind or destroy the engineers' efforts to clean up the air. Ecology is us and we are all part of the problem.

Carburetors supplied by Chevrolet have automatic chokes to provide easy starting and smooth warm-ups for the engine. These have fast-idle mechanisms to keep the engine running a little fast until the first time that you open the throttle after the engine has a bit of heat in it. Some big-block engines have a device which pre-heats the carburetor air. This improves engine performance when the engine is cold, but after the engine warms up, the system, like all other smog devices, reduces performance. The system provides only heated air to the carburetor inlet until the air temperature reaches $85^\circ F$. When underhood temperatures exceed $128^\circ F$, the system provides underhood air which is not preheated by the exhaust-manifold stove. When it is time to race, you'll want *fresh cool air* directed to the carburetor.

BASICS OF CARBURETION

The carburetor is a mixing and metering device which automatically combines air and gasoline in the correct proportions for varying engine speeds and loads. Although it performs a complex job, the carburetor is a simple low-cost device.

A carburetor is a restriction (venturi) in the engine's air-inlet path to measure air and create a reduced-pressure area. Discharge nozzles in this reduced-pressure area connect to a float bowl which is vented to atmospheric pressure. Although the nozzle is higher than the gasoline level in the bowl, a predetermined decrease in pressure at the nozzle causes fuel to flow through it. As air is sucked through the carb by the piston strokes, fuel is added to the air stream in relation to air speed through the venturi.

At speeds above idle when fuel is flowing from the discharge nozzle, suddenly opening the throttle at low RPM instantaneously reduces the air velocity so that there is not enough signal to draw fuel from the nozzle. The mixture leans out. Fuel also deposits on the manifold walls, further leaning the mixture. The accelerator pump covers up this lean period by mechanically injecting fuel.

Good low-end punch with an excellent mid-range and a better-than-average top speed is provided by a carburetor system giving a minimum pressure drop not greater than 3.0 inches of mercury (Hg) at WOT (wide-open throttle). This is approximately what you get with the stock carburetion system. Racing systems are set up to give 1.0 inch — or even lower pressure drop in some cases — under full load at peak RPM and wide-open throttle (WOT). The pressure in the manifold is approaching atmospheric under these conditions.

This is not ideal for your street-driven car!

When you add carburetion capacity to get in this ball park you add more problems than performance. This is because adding carburetors or increasing venturi size decreases air velocity through the carburetors. When velocity falls below a "critical" value at which fuel and air cease to meter in the correct proportions, the engine is over-carbureted and produces little power at or near the RPM where the critical value is reached.

Arrows indicate mixture-distribution correctors on Chevy's version of the 850 CFM double-pumper (3955205/List 4296). The tabs keep cylinders 4 & 5 from running lean at WOT when using the stock high-rise manifold without a plenum divider. A different tab arrangement on the Holley 3310-1 aftermarket version of the 780 CFM carb accommodates the mixture-distribution idiosyncrasies of the stock high-rise manifold with divided plenum.

Undesirable results include: poor low-speed torque, weak part-throttle mid-range, and a powerful top end once RPM is increased to the point where air velocity will cause fuel and air to meter correctly. The real loss for the street-runner is in flexibility of engine operation over a broad RPM range. Remember — the larger the venturi, the greater the top-end capability — but with a resultant loss in low-end response. Engines with a lot of venturi area . . . even large-displacement engines such as the big-block . . . can be said to have a "soggy" bottom end.

The problem is partially solved on your big-block Chevy engine by the four-barrel carb which has both primary and secondary venturis. These provide additional carburetion in variable fashion so that the engine is neither overcarbureted at low speeds, nor undercarbureted to a great extent at high speeds. The primaries provide the air/fuel mixture for idling and mid-range—and two secondary venturis come in for top RPM. The secondaries, incidentally, are actuated by air flow through the carburetor in the case of the Q-Jet and "vacuum-operated" Holleys. A few big blocks are equipped with Holleys with mechanically operated secondaries.

While the carburetor is off of your manifold, look at the area under the carb base. There is probably a divider under the base of the carb. If you have an engine which came from the factory as a very-high-HP engine, the manifold may not have this divider. So what is it, what's it for—and why do some manifolds have it and others don't? An explanation from Vic Edelbrock can help our understanding. "The divider separates the two planes of a 180°-style or two-plane manifold. The full divider isolates the two volumes of mixture in the manifold. These are alternately activated—first one plane and then the other—as the intake valves of the cylinders open. Each of these mixture volumes requires energy to be activated during intake-valve operation. Removing the plenum divider combines the two planes and increases the volume of mixture which must be activated each time that an intake valve opens. In terms of throttle response and mid-RPM torque, removal of the divider disrupts mixture velocity in this engine-speed range, leading to soggy response and poor off-the-line acceleration. It is simply a matter of more mixture volume requiring more energy to be activated. However, for engines normally operated above 4000 RPM, removing the divider lets the engine "see" a little more flow capacity at high RPM."

Edelbrock goes on to say, "In a sense, the divider can be used as a *fine trim tuning tool* by gradual removal of material from the divider until optimum performance is obtained for the desired RPM range. We suggest that the dividers be left undisturbed for all but long-cammed maximum-effort engines which require little—if any—low- and mid-range torque and throttle response." Incidentally, manifolds without the divider are termed "open-plenum" and manifolds with the divider are called "divided-plenum." Just remember that an open plenum "cushions" pulsations in the manifold so that they are not so harsh. The larger the plenum chamber volume, the more the softening effect. As the pulses are softened, the incoming air stream does not pull as much fuel from the discharge nozzles, so the mixture leans out in the mid-range. The cure is often excessively rich jetting so that the mixture will be correct in the mid-range, which richens the mixture at all RPM. This gives bad mileage and "soggy" mid-range and "boggy" off-the-line performance unless RPM is high so as to "launch" the car. You don't want these characteristics, so make such changes with the power valve.

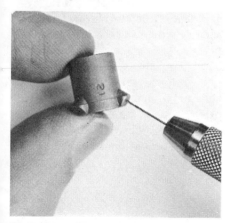

Tuning accelerator-pump "shooter" size is detailed in HPBook's "Holley Carburetors & Manifolds."

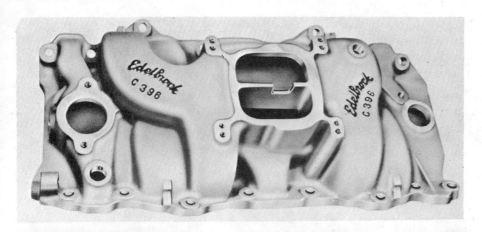

Two popular Edelbrock big-block manifolds. Top is the C-396 for small-port (oval) heads. A similar manifold (C-427) is available for the large-port (rectangular) heads. Lower photo is the very effective Tarantula design which is also offered in small- and large-port versions. Unit shown here was being modified in plenum chamber to work with 1-11/16-inch throttle bores (or a NASCAR restrictor plate). If the class rules require such restrictions, be sure to check with Edelbrock for the details of what must be done to the manifold to get the best performance.

This is the 1971 version of the Edelbrock competition Tunnel Ram manifold mounted by two 750 CFM double-pumper Holley four-barrel carbs. This is an excellent manifold for drag strip use, but is misery itself for street use—anything under 4000 RPM and you're in trouble!

Adding carburetion capacity (more carburetors or larger venturis) causes flat spots at low speeds and reduces engine flexibility. You can't argue with the laws of physics—with the same engine speed and displacement, a large venturi area slows down the mixture and detracts from good mixing of the air and fuel. The engine will be hard to start and it may be almost impossible to get a smooth transition from idle to the mid-range. These problems won't bother a drag racer or road racer, but they can make a street machine extremely tough to live with for normal driving. A long-duration camshaft further complicates the situation by pumping back a part of the intake charge, further reducing the average air speed through the venturi.

Driving an engine with too much breathing capacity is work, not fun! It will stumble, cough and spit until 3000 RPM or so—then it'll really hustle to 7000 or so. You may not need *or even want* a 7000-RPM engine. You could be more interested in acceleration than top speed. The most important factor for acceleration is torque, not peak HP. If you need drag strip speeds in the neighborhood of 90 to 110 mph, concentrate on torque—which will take just one four-barrel carb for most big-block engines. A pair of Holleys—any size—on a tunnel ram may look impressively sexy or boss to you and your friends, but they'll destroy torque—which is what you need to get to the finish line ahead of that fellow in the other lane.

Don't buy the biggest carb that you can find. Chances are that your engine will run best with a 650, 780, 800 or 850 CFM Holley. Whatever you do, do not invest in the three-barrel 950 CFM Holley—or any larger carburetor—unless you are absolutely certain that this is what you really need to get your engine to run in the RPM range that your application demands. This is so much carburetion that we cannot recommend one for any street or street/strip application. The man who buys only one carburetor and figures that he has the ultimate "hot setup" for his car is fooling himself. It is better to experiment with several carbs with different air-flow capabilities and let actual performance tell you which one is right for your job. A small carb often provides better drag ET's—and a lot more pleasure in driving—at the same time. Work with your friends who'll cooperate by trading carbs for a few test runs.

Quickly open the throttles of any carburetor arrangement with large venturi area and the engine will misfire momentarily—or even die! Why? Because manifold pressure rises almost to atmospheric, air velocity drops . . . and fuel deposits on the cylinder walls. The mixture, now mostly air, arrives at the cylinder too lean to support combustion. Many auto enthusiasts have the idea that quick throttle opening floods the engine with excess fuel, thereby causing the bog, stumble or roughness. 'Tain't so!

The problems are worsened in big blocks with large-port heads and big intake valves. This is why experienced tuners direct their customers toward the use of a small carburetor with a small-port manifold and small-port heads . . . especially for exciting street performance. And, if you are buying a carburetor, you may want to consider one of the "double-pumper" Holleys which are set up to ensure that the engine gets a massive dose of fuel to help overcome the bog which normally occurs when you "jump on the throttle."

Holley engineer Don Gonyou says, "The wilder the cam and the bigger the carburetor — the more 'hole' or bog that must be covered up, hence the trend to larger accelerator pumps — such as the Reo type which we use — or dual-pumper carbs with two complete accelerator-pump systems."

Top and side views of factory three-carb (3x2) manifold as used on high-horsepower Corvette engines through 1969. Center carb is primarily used for cruising. End carbs are vacuum-operated secondaries which open on demand. All carbs are Holleys. The ancient hotrod image required multiple carbs for the "go-fast" look—but Holley's efficient four barrels changed that! Although a large-port manifold is shown here, small-port three-carb manifolds were also supplied on some big-block engines.

This air-cleaner base is used on 1967-71 Corvettes equipped with the L-88 fresh-air hood. To duplicate it, you'd need: air-cleaner base 6422188, gasket 3919812, screen 3902396, cup retainer 3902394, diffuser 6423906 and wing nut 219281. You'll also need seal 3934165 if you actually use a fresh-air hood.

This photo of two aluminum hi-riser manifolds for Holley four-barrel carburetors clearly shows the open plenum of 3947083, as compared to the fully divided plenum under the carb in 3947084. Both are machined from the same casting 3933163, so the only way to tell them apart is to look at the area under the carb. Do not buy an open-plenum hi-riser for a car that is to be driven on the street. Both manifolds are large-port designs.

NOTE: When checking for full-throttle, do not merely actuate the linkage at the carburetor by hand. Looseness and wear in the linkage or cable will not show up unless you include the entire system—which requires mashing on the foot pedal.

Holley's GPH-110 hi-perf fuel pump with remote pressure regulator is the pump to use if you are going to buy an electric one. It is the highest capacity pump on the market and is recommended for the pro racer. It has 3/8 pipe fittings. The positive-displacement vane design has a sealed dry system. That is, the fuel does not flow through the motor as on some pulsating designs.

Some hi-rise-type manifolds require additions be made to the hood for clearance. This box is open at the back because there is a high-pressure area at the base of the windshield.

IMPROVING STOCK CARBURETION

There are many reasons why you might want to improve stock carburetion, especially if you have a Q-Jet. You may want to contend in "stock" classes at either drags or autocross races—or both. Or, you could want every last ounce of performance which you paid for in the stock engine. Stock carburetion systems are fine for economy and all-around uninspiring performance. The small-venturi carburetor gives good low-speed torque, fine engine flexibility and restricted power output at high engine speeds. Simple changes to well-engineered stock-carb systems can provide enormous performance improvements, but you must keep in mind the fact that every "advantage" has an offsetting "disadvantage."

Careful tuning of the stock carb will usually add HP, but it's hard to find these missing horses unless you have a mixture analyzer or a dynamometer . . . or both.

The first thing you should do—even if you never plan to race—is to remove the air cleaner (temporarily!) so that you can peer into the carb's air-inlet horn. Have someone else mash the pedal to the floor as you check with a flashlight to make sure that the throttles fully open (not slightly angled). If they're not opening fully, figure out why and fix the problem. Any time you remove and replace the carburetor—check again to ensure that you have a fully opening throttle. It is the easiest thing to overlook and the cause of a lot of lost races or poor times. Any honest racing mechanic will admit that he's been tripped up by a part-opening throttle *at least once*.

Now put the air cleaner back in place. Don't leave it off. This is a temptation that you must overcome. The time-honored mark of the performance-oriented owner may be an engine with no air cleaner, but you can't afford to copy all of the things which have been proved not to work. You must proceed in a heads-up intelligent fashion to keep everything working for you—not against you. The engine needs all the help you can give.

The air-cleaner directs the air into the carburetor in such a way that the vents work correctly and so that the air gets into the air bleed jets in the correct fashion. A carburetor engineer can use almost any carburetor without the air cleaner or its base, but you may never acquire the specialized knowledge that is required. So, rather than work against yourself, use what has been proved to work successfully and you will be that much farther ahead of your competition. You have probably read a lot of articles which said to be sure to leave the air-cleaner base in place, even if you had to remove the air cleaner for some obscure reason. Lest you think that the writers have been kidding you, one dyno test series showed a 3 HP loss by removing the air-cleaner base. This little item causes air to flow into the carburetor with less turbulence. It is essential.

Air cleaners also protect against fires caused by starting "belch-backs"—reduce intake noise—and reduce engine wear. Intake noise can be horrendous—even worse than exhaust noise—on a big block that's turning up a lot of RPM.

Before leaving the subject of air cleaners, let us remind you to look at the way the stock cleaner is designed before you invest in some flat-topped short cleaner because it looks good. Note that the high-performance cleaner stands high above the carburetor air inlet. This design allows adequate space for the incoming air to enter the carb correctly and with minimum turbulence. With a flat filter sitting right on top of the air inlet you may get HP loss.

Naturally, the next thing to do is to bring the carburetor or carburetors up to the latest-year specifications as shown in the Service Manual. Of course, you'll want to make sure that the carburetor is clean and that the float level is correctly set.

In general, the main jet is correct for maximum performance at WOT. The latest models with low emissions are jetted so lean at part throttle that the engine runs hotter than would otherwise be necessary. Power valves are carefully

Mixture checking by plug reading: page 49

Top and side views of the low-rise big-port aluminum manifold 3977609 supplied on the 450 HP 454 for use with the Holley four barrel. Here you see the direct comparison with the low-rise small-port cast-iron manifold 3977608 used on 390 HP and lower performance 454's. The cast-iron manifold can be equipped with a spread-bore Holley four-barrel. Both manifolds have a sheet-metal shield directly under the heat passage to avoid exhaust heating the oil which is splashed under the manifold. A 3931093 inlet-manifold oil shield should also be used under the manifold as an additional heat barrier so that the "cold" manifold will not be heated by the oil when the exhaust heat riser is blocked off.

selected to give best part-throttle performance with low emissions and best economy while maintaining best WOT mixture. If you make the mixture rich enough for best mid-range and part-throttle performance by changing the power valve channel restrictions, mileage will suffer.

Changes which can affect carburetion in this manner include larger valve sizes, bigger displacement, ignition timing, special camshaft, changes in compression ratio, elimination of intake manifold exhaust heat, etc. An efficient exhaust system greatly affects the engine's overall breathing ability. Correctly exhausted cylinders have less residual burnt charge and can handle more intake charge—supplied automatically by the carburetor as added air is sucked through the venturis. The only time that jet changes are needed is when an exhaust system changes the intake-system pulsing seen by the carburetor—or when the carburetor is too small for the application.

You may be tempted to rip out the choke(s) until you look at the carb to see that there is so much area around the choke—as compared to the venturis being supplied—that there's no way to get more air through the carb by taking out the choke. Taking out the choke causes further wasted effort because then you still have to plug the choke-shaft holes so that all air entering the carb will come through the air filter.

Improvement *is* available by cooling the mixture, assuming that you are sweltering in summer's heat—because this is something that you don't want to do in wintertime or you'll get carburetor icing. Eliminating the exhaust flow through the heat-riser passage is accomplished by cutting pieces of thin (0.020-0.030-inch) stainless steel to block the passage. Remove the manifold and slip the metal in on each side to block exhaust flow across the underside of the manifold. Use new gaskets to minimize the possibility of coolant leaks. Dyno tests show that this simple trick adds several HP with no other changes. It also increases the time required for engine warm-up. Some aftermarket companies offer gaskets for LPG conversions with the heat-riser passages already blocked.

To cool the mixture even further, use a "cool can." These standard items for the gung-ho drag racer use tubing coiled in a container of ice or dry ice to cool the fuel before it gets to the carburetor. The fuel line is merely cut apart so that the cool can can be inserted as a section of the line. The HP improvement is minimal, but every little bit counts in drag racing. There are lots of different fuel-cooling cans on the market.

If you live in "hot country" where summer temperatures go out of sight, consider installing a fuel radiator. You can run the fuel line through a Corvair oil cooler which you'll mount up in the airstream at the front of the car. Such a device will practically eliminate any vapor-lock problems and adds both mileage and performance. The radiator will not provide the cooling which can be had from a cool-can device, but is helpful for road-racing applications and ordinary highway driving.

Holley Carburetors & Manifolds tells you everything you need to know to select, install and maintain any Holley carburetor. The performance-tuning section is a must for any serious racer.

Underhood temperatures can go out of sight on large-block installations—and HP goes down as a result. This owner has installed the 70-71 heat-insulating aluminum shield under his Holley to keep some of the manifold heat off of the carb. A good idea! Ask for P/N 3969835. Chevy sells these for Q-Jets, 3982944—'bout two bucks a copy. Braided-steel covers on plug wires indicate that this plastic car is FM-equipped. Tin box around distributor is a 'Vette trademark. Chrome rocker covers could be sandblasted and painted flat black to help engine get rid of its heat easier.

BROADENING THE TORQUE CURVE

Development testing at Chevrolet engineering proved that the torque curve of the fuel-injected aluminum engine can be smoothed out to eliminate the characteristic peaks and holes of a ram-tuned intake. The drawing shows unequal-length ram tubes and their locations and dimensions for use in the Crower/McKay 2.9-inch-butterfly injection manifold. Use of the ram-tube lengths shown yields a virtually smooth torque curve from 3500–7000 RPM with any displacement from 427–500 CID.

This configuration has been adequately tested for durability and shows no detrimental effects on the engine. Can Am engine builders have been using the arrangement with good success since 1971.

Other dynamometer testing showed that cylinders 2,3,6,7 require less spark advance for maximum power than the other four cylinders. This can be accomplished by using Champion N-62-R (or equivalent) side-gap spark plugs in these cylinders and N-60-Y (or equivalent) extended-tip plugs in 1,4,5,8. The same 38–40° maximum spark advance is still used. This plug recommendation is good for all big blocks, whether injected or carbureted — and for both iron and aluminum blocks.

The unequal-length recommendation was arrived at by running the engine as two separate four-cylinder engines to determine the best intake systems for each of the port configurations. (The ports opening into the center of the cylinder work better than those which point into the cylinder wall.) A different length is required for ram tubes used on other injector systems with smaller openings.

Edelbrock took advantage of this information to change the design of the tunnel-ram-type manifold for improved performance — as shown in an accompanying photograph.

Using these recommendations for drag-type engines should also be helpful in improving performance.

Exhaust-system changes are not required when using these recommendations. Equal-diameter pipes seem to work o.k. due to the fact that the exhaust ports all work equally well.

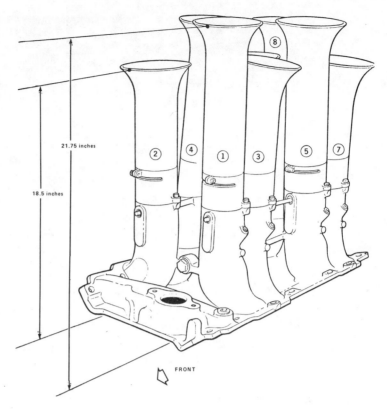

Two different ram-tube lengths are specified by Chevrolet Engineering to flatten out the torque curve on fuel-injected Can Am engines. Lengths are for 2.9-inch-throttle injectors. Different lengths are required when other throttle dimensions are used. Technique can be used for injector-equipped drag cars and boats. Chevy drag-engine builders have been overlooking the idea of a broad torque range in favor of all-out HP. This staggered-ram idea could be coupled with an intelligent use of roller-tappet cams to really wake up the big-block. It's certainly worth thinking about! The idea has certainly caught on with the Can Am builders.

First Edelbrock TR-2X manifolds (Tunnel-Ram) had rectangular runners with equal dimensions (top). Later TR-2X had its runners spread apart. One runner was raised and one lowered (in each pair) and one of the runners had its cross-sectional area (and therefore, volume) reduced to take advantage of the intake-system tuning phenomena indicated in Chevy's unequal ram-tube-length recommendations.

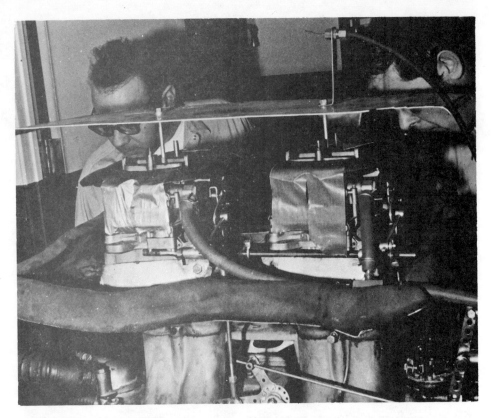

Foam rubber and aluminum baffle under carbs snugs up against underside of Bill Jenkins' racecar hood to keep down air temperature entering carbs. Check fuel-line size. Note carb air-cleaner bases taped to top of each carb and flat plate to provide correct air flow into carbs. The name of the game is brute horsepower . . . any way you can get it . . . within the rules, of course.

Bell-shaped entry works about as well as an L-88 air-cleaner base to straighten air flow so air vents and bleeds work correctly. Screw in secondary lever (arrow) is not for street or normal street/strip operation. If you've done this to your street-driven machine, take out the screw and reinstall the vacuum actuation to improve performance and economy.

RAM AIR AND COLD-AIR KITS

Some would-be speed tuners have the mistaken idea that they can add a mild form of supercharging at low cost by connecting rubber hoses to their carburetor(s) from scoops on the hood or under the front bumper. Any apparent improvement thereby obtained is due to the engine receiving cold air—*not from any supercharging or "ramming."* Ram air can provide minor HP improvement at very high speed: +1.2% at 100 MPH, +2.7% at 150 MPH, and 4.8% at 200 MPH. But, according to Colin Campbell, ramming to get this small power addition also creates problems. Unless the ram air is fed to all parts of the carburetors, including the float bowls and all vents—carburetion is disturbed, especially at low speeds. His book, *The Sports Car—Its Design & Performance*, states that a cold-air box arranged to feed the carburetors should be designed without ram effect by allowing free flow of the air out of the box feeding the carburetor filtered intakes.

Cold air will give you more improvement than ram air because 1% HP is gained for every 10°F. temperature drop. This assumes that the mixture is adjusted to allow for the change in mixture density and that there is no detonation or other problems. Cold-air-box use is climate-limited because too-cold temperature will cause icing. Air scoops are o.k., provided that they are not connected to the carburetors—but to cold-air boxes or air cleaners which are vented correctly.

For complete details on ram tuning intake systems, refer to Philip H. Smith's book, *The Scientific Design of Exhaust & Intake Systems.* Perhaps one of the best single articles written on the subject was by Roger Huntington in the July 1960 HOTROD Magazine, "That Crazy Manifold." July and August 1964 HOTROD Magazines had two further articles which were written by Dr. Gordon H. Blair, Ph.D. All are worth reading, however, the use of a dyno to measure the real performance of a specially constructed "tuned" system—either intake or exhaust—is absolutely essential.

If you are running a cold-air kit which picks up cold air at the front of the car, be prepared to change the air-cleaner filter element at regular intervals, perhaps as often as once a week on the street in dustier areas. Front air inlets look "boss," but they are giant vacuum-cleaner entries to the carburetor. You'll be better off ducting cold air from the cowl just ahead of the windshield, using GM's standard parts for the task. That area still gets airborne dust, but it is several feet off of the ground—away from the grimy grit encountered at road level. If you like a car that looks "stock," the use of fresh air from the cowl is another way to get performance without making the car look like a racer.

SINGLE-CARB MANIFOLDS

When you are looking for a high-performance single-carb manifold for your big-block Chevy, you've got a lot to choose from: Edelbrock, Holley and Weiand. These are all open-plenum designs, with the plenum kept to a reasonable size to retain high mixture velocity at low and mid-range RPM, especially on the street versions.

The latest of these manifolds was introduced in 1976 by Holley Carburetors. They are called *Street Dominator* and *Strip Dominator*. Edelbrock calls their units *Tarantula* and *Scorpion*. Weiand has *Exterminator*. Street types typically have exhaust heat and are drilled for both square-bore and spread-bore flange carburetors. Street units have provisions for mounting the factory EGR valves.

Competition versions of these manifolds are usually equipped only for square-bore carburetors, have no exhaust heat, and there is no provision for smog equipment.

Edelbrock manifolds may have an anti-reversion step or mismatch where the manifold meets the cylinder head. That is, the manifold ports do not match with the cylinder head ports. The general consensus among performance-oriented engine builders is to grind out this step so the ports in the manifold match those in the cylinder head. This is probably not as important for a street engine as for an all-out racing engine.

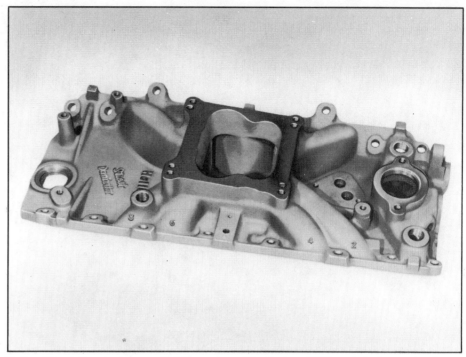

Holley Street Dominator manifold is equipped for factory EGR and choke mechanisms. Carburetor flange accepts either spread-bore or square-flange carburetors. Exhaust heat is provided for fast warm-ups and smooth running in cold weather.

Holley Strip Dominator manifold is pure competition. Isolated-runner design ensures coolest possible mixture gets to the cylinders. There are no EGR, choke or exhaust-provisions.

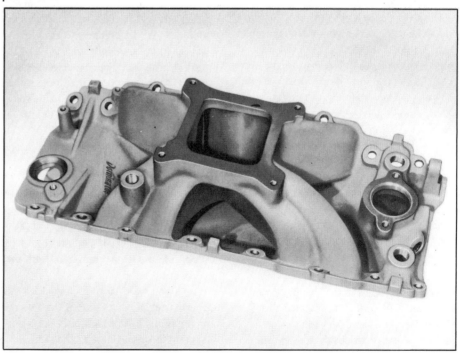

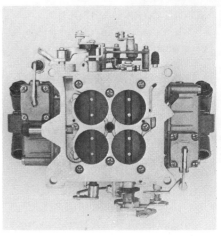

Double-pumper carburetors have revolutionized high-performance carburetion since their introduction in 1967. Pumps for both primary and secondary throttle bores provide adequate fuel to ensure that no bogging will occur when the throttles are opened at an RPM too low to provide good velocity. These are available with flow capacities from 600 to 850 CFM.

Small-primary/large-secondary throttle-bore concept is generally conceded to be the plan for future carburetors, whether high-performance-oriented or not. Holley's 4165 series was the first spread-bore design combining high-performance capabilities with the ability to provide specification emission performance at the same time. Note the two accelerator pumps and the mechanical secondary linkage. These are designed as bolt-on replacements for your big-block.

HOLLEY CARBURETORS
Which one?

Always use the smallest carburetor which gives the best performance. If your engine came equipped with a Holley, there's a good chance that you will want to go right on using it, once you have identified which one it is. However, if you are going to buy a Holley carburetor, you should have a good idea of what you really *need* before rushing to the speed shop or Chevy dealer with your checkbook in hand. You might think that you want a really big Holley, such as the 1050 or 1150 CFM 4-barrel Model 4500. Impressive as they look, you really wouldn't be happy with either of these—and certainly not two—unless you were off to battle the Pro Stockers.

Chances are that your best buy, especially if you have to drive the car on the street, will be the old favorite: the Holley 3310-1 780 CFM with vacuum-operated secondaries. Cars will often run faster with a vacuum-secondary carb than with the double-pumpers because the engine gets the right amount of air and fuel in relation to engine speed and load. We like Honest Charley's description of the Holley 3310-1 as "The most dependable, runniest carb ever built." At the low price that these sell for, we'd recommend bypassing the more expensive double-pumper 850 unless you *really* have a need for it. The 780 typically costs $150 or so at your local Chevy or Holley dealer.

When using the 780 or 800 CFM Holley on the divided-plenum manifold, use 72's in the primaries and 76/78 in the secondaries.

If you are running two carbs, as on a tunnel-ram manifold, use the carbs specified by the manifold maker. But, if you want to go racing with a single four-barrel, consider one of the "double-pumper" Holleys which only feeds an accelerator-pump shot to the secondaries when they really need it.

Get the aftermarket models

When buying a Holley, get yours as an aftermarket equivalent. If you buy from a Chevy dealer, avoid carbs with a GM part number. Naturally, there is an exception to this. For oval-track, long-distance or boat use, the open-plenum Chevy manifold operates best when used with the Chevy version of the 850 Holley. This carb has mixture tabs built onto the discharge nozzles. The result is a more even mixture distribution which is critical with this manifold — even if stagger jetting is used to richen up cylinders four and five. If you're handy, mixture tabs can be fabricated from Devcon "F" aluminum epoxy right on the booster venturis of an aftermarket carb. Look at a Chevy carb to see what needs to be done. Economy is improved when the spread-bore type Holley (Q-Jet replacement) is used because these have smaller primaries. However, the improvement will only be there if you drive so that you are not always causing the secondaries to operate.

Separate pumps are a help with progressive secondaries because there's no pump shot ahead of the time when the secondaries are to be used. With 1:1 non-progressive secondaries, two pumps have more capacity to cover holes, therefore allowing the use of a larger carburetor than might otherwise be possible.

Chevy was the original manufacturer buying the double-pumper and thus was first on the scene with these carbs on their machines. However, as of 1970, Holley had the dual pumpers available in their aftermarket line for sale to anyone who wanted it—including walk-in buyers at speed shops.

If you plan to race a 427 or 454, good throttle response and smooth idle can be provided by one 850 CFM Holley. This carburetor is used on many of the late-model, large-block Corvettes and is available from Chevy dealers under part number 3955205, or Holley 3418. It is not an economy carb! And, it may be too big for your street-driven 396 or 402.

This Holley has 1-3/4-inch-diameter throttle bores, mechanically actuated secondaries and two accelerator pumps . . . one for the secondary barrels. If this carb is used with the open-plenum manifold, stagger jetting will allow an equal distribution of fuel to all cylinders. Although jetting may vary slightly because of altitude, engine use, etc., it should be similar to: left front No. 80, right front No. 76, left rear No. 76, right rear No. 78. To richen or lean the mixture, keep the same basic balance as you change jets up or down.

If the 850 is used on the divided-plenum manifold, remove the mixture-distribution tabs from the boost venturis in the left front and right-rear barrels. Because these are the only ones which have these tabs, you'll spot them immediately. They are 1/8-inch-thick parts on the booster venturis. Once this has been done, install 74's in the primaries and 78's in the secondaries.

Double-pumpers, it should be noted, can murder economy because that pump covers up the "hole" caused by banging the secondary open

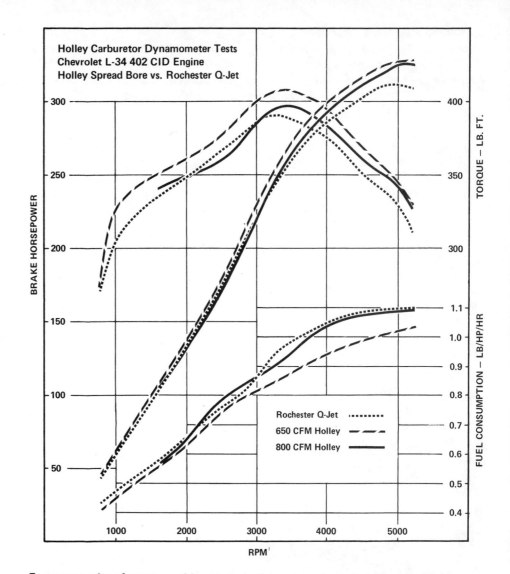

Economy and performance seldom come in the same package, but the spread-bore Holley may be an exception from what we can read in these dyno charts. The 650 CFM put out more peak HP and torque—on less fuel, but above 5000 RPM the picture will change. One of these carbs will drop right onto your Q-Jet manifold.

when the engine may not be able to use it. You get a double shot of fuel every time you mash the throttle—and there goes your mileage, friend.

More about tuning the Holley

Stay close to the stock jetting arrangement. If you bore or stroke your engine, the jets do not necessarily need to be changed. If you have to use jets which are more than two or three jet numbers away from stock, the chances are very good that you have done something wrong. Or, something could be amiss in the engine.

More time is wasted playing with vacuum-diaphragm springs than you can possibly imagine unless "you've been there." Nearly every would-be racer starts out with the idea that the vacuum opening of the secondaries is hurting his car's performance. Just the contrary may be true—slow opening of the secondaries so that no bog occurs when they come open may easily add speed and cut ET's at the drags. Opening them too quick may make for slower times. Don't approach this subject with your head in the sand, keep that stock spring that you take out when you buy others to try—so you can put the stock one back when the others fail to give you improved times at the drags. The lighter springs seldom improve acceleration times.

Vacuum-diaphragm springs for the Holley vary from high-tension (purple) through medium (yellow) to low (white). The lower the tension, the quicker the secondaries come in—and the more chance that you'll create an awful bog. Most engines run best with the springs supplied in the carburetor.

There are so many things that you can do to tune a carburetor that there's no way to cover it all as part of a book like this one. One of the things that Holley does is to provide pump "shooters" or discharge nozzles with different orifice sizes. These typically vary from 0.021 through 0.031 inch. Accelerator-pump-discharge time can be tuned by varying the size of the shooter orifices. With mechanically operated secondaries (non-progressive) especially, it's sometimes necessary to open the orifices to as much as 0.040 and add one of the large-volume Reo-type accelerator pumps. These, as you may already know, may require raising the carburetor 1/4 inch for manifold clearance.

Nothing mysterious about a cool-can! Fuel in coiled metal line cools off because the can is packed full of ice or a combination of dry ice and alcohol to really lower the temperature and avoid any tendency of the fuel to flash into vapor as it enters the carburetor bowl. These are standard wear on all drag cars during the summer, especially.

In a drag-race situation, excessively rich jets are sometimes used because they *are* needed. If 80's are needed on the dyno, then you may have to go to 82 or 84 on the strip for best acceleration.

Double-pumper carbs help to reduce or eliminate this tendency and thus are quite popular for drag racing, even though they may murder economy on the street and highway. Having plenty of fuel available at the carburetor with large fuel lines and adequate pressure helps, of course.

Holley's accelerator pumps are actuated by cams which can be changed to modify their action, or moved on the lever to provide a richer shot for cold-weather operation. This is another area where you can experiment if you have the time and inclination. Just remember that there must be 0.015-inch clearance between the actuating lever and the fuel-pump rocker lever at WOT, or you'll distort the diaphragm and perhaps even tear it at full travel. No clearance is wanted at closed throttle.

Power valves are used in most Holleys. Many racers replace them with a solid plug, but this is not the correct way to tune because numerous power valves with varying characteristic are available to tune the carb for almost any application. A great deal of fine tuning can be done with the power valve, so don't be in such a hurry to discard it. And, if you take it out, you'll have to rejet to compensate for its removal. On a drag-race car only, plugging the secondary power valve may help eliminate a stumble at the starting line or just after a shift caused by the slosh of fuel in the secondary float bowl which can uncover the valve. Never plug the primary power valve for a street-driven car. A lot of details on performance tuning Holleys is in HPBooks' volume, *Holley Carburetors & Manifolds*. This fact-filled 192-page book is a must for your automotive library.

Some Holleys are equipped with white-plastic vent baffles, called "whistles," to prevent fuel from sloshing out through vent tubes under hard acceleration. If you install these in your center-pivot float bowls, you may have to shorten the whistles and secure them with epoxy so that they will not prevent correct operation of the floats.

Whistles eliminate the need for extending vents with rubber hose, etc.

Holleys have two styles of float bowls. The center-pivot type is best for racing where side loads are imposed on the car. The cheaper side-hung type is often used where two carbs must be jammed closely together, as on a tunnel-ram manifold. Unless you are autocrossing or road racing, the advantages are that the float levels can be adjusted externally and, you can install different needles and seats—up to 0.110 or 0.120 inch—without removing the float bowls. If you have a choice, take the center-pivot type, but realize that your original stock air cleaner may not fit onto the carb which has center-pivot floats. This is the case with the spread-bore Holley 800 with the center-pivot bowls.

Just inside the fuel-inlet nut at each end of the Holley carb is a sintered bronze fuel filter. These should be removed and discarded. A large paper-element fuel filter should then be installed between the fuel pump and the carburetor.

QUADRAJET

The Q-Jet carburetor comes stock on a vast number of large-block Chevys. It is a relatively sophisticated unit which is designed for smoothness, power, flexibility and long, troublefree operation. The carb contains four barrels: two primaries and two secondaries. The primaries are unusually small for excellent low-speed torque and good mileage. Secondary throttle bores are big. The centrally located fuel bowl gives a compact design and more importantly helps to eliminate fuel starvation/slosh problems encountered in hi-perf cornering situations.

If you are a hard-charging driver interested in competition, the Q-Jet is probably too small for use on anything other than a modestly modified 366/396/402 with small-port heads. So, our comments about improving the carb should be regarded as the best ways to get more out of something which is too small to begin with . . . if you are going racing. The Q-Jet flows about 700 CFM.

Until 1971 when Holley introduced *their* spread-bore carb, the big-blocker had a real problem when changing to a Holley for more air flow or easier tuning. A new stock or speed-type manifold had to be purchased to get the correct carb base spacing—unless

Pen point indicates air-valve adjustment screw. Arrow points to socket set screw which secures adjustment.

an adapter plate was used—which knowledgeable owners avoid. The wallet-busting price of a carb and a manifold caused most owners to stick with the Q-Jet, even though it provided less than optimum performance for competition use. Now that the Holley is available, you may want to proceed directly to the store to buy one, especially after reading all about what's necessary to get the Q-Jet tuned up.

To work on the Q-Jet with any degree of finesse requires following instructions in the Chassis Service or Overhaul Manuals for your particular model. Otherwise, trying to make it work—especially on the 70-and-later models with all that emission-control stuff—will be like trying to find your way out of a dismal swamp on a moonless night without getting eaten by alligators or bitten by snakes. Manuals are cheaper than new carburetors every time. It's

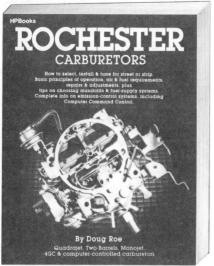

Rochester Carburetors is the definitive book on this make. If you are running one of these carbs and want the utmost in performance—or best economy, this book helps you solve problems you didn't know existed.

amazing how many neophyte tuners will read a magazine article with all kinds of tricks and believe that this is really what the man did to his carburetor. So, they start drilling holes, filing things and tightening springs, etc.—getting the part into a condition so far from stock that a trained carb man may not be able to repair the damage without a lot of new parts. "A little knowledge is a dangerous thing."

Other than a good basic rebuild by carefully following the instructions in the kit *and* in the factory manual/s, a few modifications may be performed in pursuit of better response. First of all, if your carb has a bronze filter in the entryway, chuck it. Paper-element filters from later carbs can be fitted in, or use an in-line fuel filter. And, don't take out the choke because you cannot help performance by doing so. You may cause the carb to run worse because the choke is part of the air-horn configuration which makes the air bleeds work correctly.

Primary metering

Take the air horn off to get at the primary metering rods and jets. Pop out the power-piston assembly with the metering rods by bouncing the assembly lightly and rapidly with your finger.

There are 46 different primary jets, ranging from 0.050 to 0.099 inch. They can be ordered through your Chevy dealer or a United Motors dealer. However, you can't drill the jets so that they will flow accurately or even similarly, so buy pairs of jets in the sizes that you want to try. Then you don't have to do the job all over again when you think that you are finished tuning. Do not attempt to install Holley jets because they will destroy the different threads in the Q-Jet body.

The first mistake made by most people in working on any carburetor is to make everything *bigger*. Leaner jetting often provides sharper acceleration. With the Q-Jet you can install smaller primary main jets if you richen up the secondary system with different metering rods so that there will be plenty of fuel to avoid leaning out when the engine is putting out power.

If you install larger primary jets, the power system and wide-open-throttle are also richened and the mid-range or part-throttle condition is also fattened up. Here is a relationship which you should note: for each 0.001" larger primary jet, install

two sizes larger metering rod; for each 0.001" smaller primary jet, install two sizes smaller metering rod. Each 0.001" primary jet change richens/leans WOT operation 0.5 A/F ratio. The rod number indicates the largest diameter of the rod portion which is fully inserted into the jet in the part-throttle position. The rods must be changed whenever you change the jets or the part-throttle operation will be excessively richened/leaned, depending on whether you installed larger/smaller primary jets.

Secondary metering

Each secondary barrel has a fixed jet. Fuel which flows through it is dependent on secondary metering rods (needles) attached to a yoke (hanger) which raises the rods out of the jets to flow more fuel as the air valve over the secondary throttles opens. The yokes are letter-coded. Each letter from "B" towards "V" drops the rod taper farther into the jet by 0.005", leaning out the entire calibration by 0.2 A/F ratio.

Secondary metering rods are numbered with two letters (AT, AX, etc.). Your local United Motors dealer's charts describe the rods' small-end diameters which control fuel flow when the secondaries and air valve are fully open. Thus, a rod change alters A/F ratio where the air valve opens and/or at maximum flow. If you want to richen the mixture at the top end, reduce the small-end diameter of the rods about 0.003" at a time by using a file or emery cloth on the rod tips as they are spun in a lathe or drill press chuck or collet. Once you have found the correct diameter by experimentation, buy the right rods from the United Motors man.

Should your engine exhibit a mid-range stumble, you may be able to cure the malady by adjusting the secondary metering yoke so that the metering rods are caused to sit higher in the jets and so that they lift higher sooner. Bending the yoke is not recommended, but you can either change the yoke or epoxy a small piece of paperclip wire under the lift arm at the pivotal point as a starter for experimentation.

When installing the air horn on the carb body, don't jam the secondary metering rods into the jets because this can affect metering.

Primary rod and jet tables: page 110

A pair of holes supply fuel to the standpipes which deliver fuel to the secondary barrels. Some tuners have added additional holes below the stock ones to add to the total fuel reserve available for the secondaries. Such holes can be plugged with lead shot to get the carb back to stock condition if the change does not help.

A few words about that air valve

Q-Jets have a giant pair of secondary throttles controlled by the accelerator. However, to avoid the bogging which is so common when a carb is quickly opened wide, both air flow and jetting of the secondaries are controlled by a unique air valve. As more air enters the carb, air gradually overcomes the spring pressure which normally holds the air valve closed. As previously mentioned, a cam attached to the air valve lifts a yoke with two tapered metering rods out of the secondary jets to increase fuel flow.

Air-valve opening point is controlled by a wind-up spring. The air valve is off-center on its shaft so that air acting on the larger area turns the spring-loaded shaft. As the air valve opens, it forms a crude V-shaped "venturi" with a sheet-metal baffle inside the secondary air horn. Fuel-discharge pipes here deliver fuel as air flows through the secondaries. The previous section described how the air valve operates the metering rods to affect the entire system.

When you open the throttle quickly, air flows immediately — but fuel lags — hence the use of an accelerator pump to add fuel to help make up for this lag. However, the secondaries do not have an accelerator pump of their own and you may want to increase the spring tension to slow secondary action. A very slight pause as you apply more throttle could indicate the need to slow the air valve by increasing spring wind-up, but not beyond one turn! Unfortunately, increased spring tension may cut down the carb's total air-flow capability because the air flow at peak RPM may not be able to open the air valve against the increased spring tension. While this may not be a problem on a 300 or 350 CID small block, the flow reduction can hurt performance on engines of 396 CID and larger.

Adjusting the spring tension against which the air valve operates causes the air valve to open either sooner or later. To adjust the spring, loosen the socket set screw under the adjustment screw and back off the screw CCW 1/4 turn at a time until you reach the zero-tension point where the air valve barely closes when you tap on the carburetor. Then, preset spring tension by turning the adjusting screw 7/8 turn CW from the zero point. Hold the adjusting screw position as you retighten the set screw. Any experimentation from this point should be 1/8 turn at a time, but the screw must not be tightened more than one turn from zero-tension or you'll distort the spring into uselessness.

A lot of owners try to quicken the air-valve opening, thereby interrupting the continuous smooth air flow that would be best for acceleration with this carburetor. Quick air-valve opening sends the car straight to Bogsville, which is located on the map a few feet down the road from where you buried your foot in the throttle. If you must experiment in this area, unwind the spring until the engine begins to "stumble" just as the car leaves the starting line, then tighten the spring a little at a time until it's eliminated — but not more than one turn.

Jim Baker, who campaigned a '68 Camaro (120.95 MPH at 11.67 seconds ET) used an elongated pin through the yoke pivot point to hold the air valves slightly open in their at rest or "closed" position. This eliminates hesitation caused by sudden opening of the air valve and may allow running closer to the stock wind-up setting. However, this kind of mod requires tailoring for the individual situation. It may not work for everyone!

Accelerator pump talk

After setting the air valve, more pump action may be required. This is especially true if the air-valve setting ends up 1/4 to 1/2 turn weaker than stock. Remove about 1/8 inch from the end of the pump-rod assembly to increase the pump stroke. This increases the volume of gasoline delivered to cover up holes caused by too quick an air-valve opening, a lopey cam, or whatever. But, do this kind of mod on an extra pump assembly so you can switch back and forth to see whether you are helping or hindering performance. There must be a little travel left in the accelerator pump at WOT or the primary will not open all the way — and neither will the secondary.

Leave the float level alone

Changing the float level to cure ills is a poor way to go. This changes all kind of characteristics in the carburetor, including where the jets begin to work and so on. Use the stock float level. If the carb needs more fuel, consider raising the fuel pressure to as much as 8 psi and install a larger needle and seat from a Buick, P/N 7023896. It has 0.136-inch diameter as compared to the stocker's 0.125 or so. The combination of a larger needle and seat and more fuel pressure may be the only way to cure a lean condition at high RPM.

Secondary opening change

Modifying the secondary lever as shown in the accompanying drawing starts secondary opening at 30° instead of 50°. If the vehicle is light or geared for quick acceleration, this can be *extremely helpful*, especially in autocrossing. This technique just about doubles the CFM capabilities at a 3-inch Hg pressure drop across the carb. But, if the secondaries are in excessively during highway cruising, economy will be *murdered*, according to sportscar carburetion expert Doug Roe.

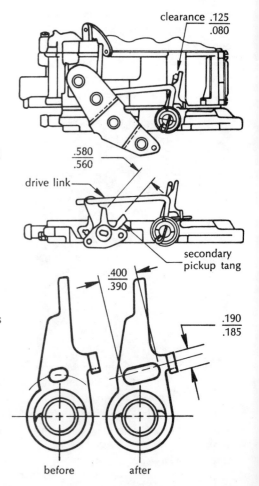

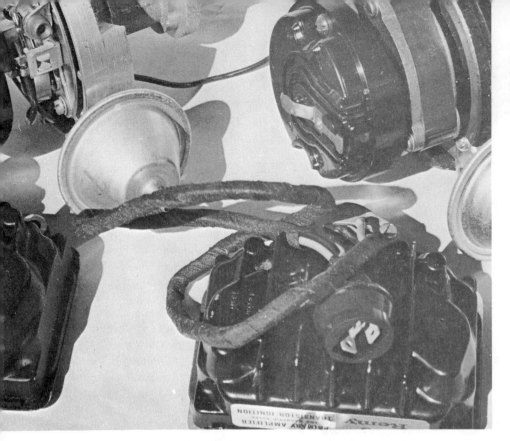

Ignition
the stock system is better than you thought

Two Chevrolet transistor ignition systems, distributors and amplifiers. Magnetic-pulse "sender" is below rotor of distributor at right. "Contact Controlled" means ordinary points . . . in distributor at left.

Best street distributor is HEI unit. The HEI system is instantly recognizable by its bulky appearance, top-mounted coil, and spark-plug type connectors. Chevy part number is 1103302.

STOCK IGNITION SYSTEM

The stock ignition system consists of the battery, a distributor, a coil, an amplifier for transistor-switched or magnetic-impulse ignitions, spark-plug cables terminating in boots over the plug connectors, and spark plugs. All of the stock distributors have one point set for the ignition system—or a toothed wheel which is surrounded by a magnetic-impulse pickup.

Other than the dual-point with tachometer drive, no point-type distributors are listed in Chevrolet heavy-duty parts list. We surmise that eliminating any need for checking point gap and ignition dwell has caused the apparent changeover to the magnetic-impulse systems for high-performance engines. With one of these systems, you can be assured that ignition timing will not change as the distributor is used. Points, on the other hand, wear at the rubbing blocks so that the gap closes and dwell angle—which determines coil saturation time—changes. Just as frustrating is the fact that this retards the ignition timing.

Chevy refers to their magnetic-impulse system as a "transistor ignition system." It features a specially-designed distributor, ignition pulse amplifier, and a special coil. Two resistance wires are used in the circuit: one as a ballast between the coil negative terminal and ground, and the other as a voltage drop for the engine run circuit. This second resistance wire is bypassed during staring. Two magnetic-impulse distributors are offered. Both have a mechanical tachometer drive. The top-of-the-line unit, 1111263, is an all-ball-bearing distributor. Because it is *strictly for racing,* it does not have a vacuum-advance mechanism. Its advance mechanism is controlled by weights which advance the toothed wheel as engine RPM increases. This is a truly maintenance-free distributor, as is the HEI described on the next page.

Next in line is the plain-bearing distributor with vacuum advance, P/N 1111267. This distributor has become a favorite with street/strip engine builders. The only maintenance recommended for this distributor's upper bushing is "at overhaul" when the plastic seal should be removed and SAE 20 oil added to the packing in the cavity. A new oil seal is required because the old one is destroyed by taking it out.

41

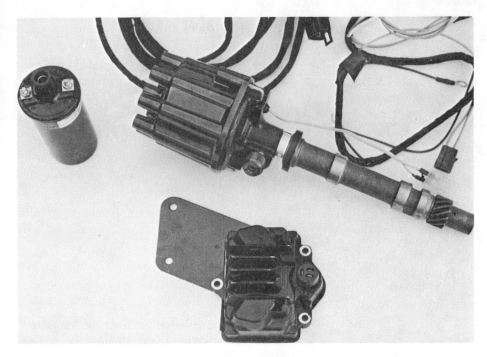

This is the good-guy distributor that Chevy sells for the real racers. It's a ball-bearing-equipped distributor with magnetic-pulse-triggered transistor ignition. All cables, amplifier and coil are parts contained in a transistor-ignition unit listed in the HD Parts List.

Delco dual-point distributor from small block can be used in big block. This one is equipped with a mechanical tachometer drive — which usually indicates that the distributor was originally designed for use with the Corvette engine. Points must be replaced with heavy-duty ones to allow high RPM operation as the stock points "sign-off" at about 6,000 RPM. This unit has a fixed point plate, hence only mechanical (centrifugal) advance. No vacuum advance can be fitted to this distributor, so consider it useful for racing only. Vacuum advance is a necessary requirement for the street machine.

Each of the magnetic-impulse distributors is used with kit 3997782 to make a complete transistor ignition system. It is a "brute-force" transistor system which incorporates a certain degree of dwell "stretching" so that the coil will produce adequate sparks at very high RPM. But, contrary to what you may have read, the Chevrolet systems supplied for the big-block and small-block engines are not capacitive-discharge systems.

Dollar for dollar, the best street high-performance ignition is GM's high-energy ignition—HEI—system. This breakerless system will put out 60,000—80,000 volts all day and fire wide plug gaps. Junkyards are full of them. They came standard on all 1975-and-later GM cars and light trucks.

A point-type distributor can be converted to the breakerless triggering system by buying two parts: 1960779 and 1964272–the rotating pole piece and the stationary pole piece with magnetic-pickup assembly. A distributor converted thusly will operate with the magnetic-pulse-triggered amplifier included in the kit mentioned in the previous paragraph. Other manufacturers make capacitive-discharge units which will operate with a stock or converted Chevy breakerless distributor. Point-type distributors can also be converted to the magnetic-pulse type by using a special kit which is described later in the chapter.

A point-type distributor with tach drive and vacuum-advance can be used in the big block (ignition parts are interchangeable between small- and big-block engines). This is P/N 1111496. Another small-block point-type distributor which is often used has dual points and a solidly mounted point plate—hence no vacuum advance. It is P/N 1110985. Or, you can add dual points to any of the Chevy V-8 distributors by installing dual-point plate 1953752.

HOW TO IMPROVE STOCK IGNITION

A few of the stock Chevy ignition-system parts are nearly always exchanged for other stock or "high-performance" parts by owners who are literally "taken in" by ads for super spark plugs, extra-special spark-plug wires, deluxe condensers, long-life points, and high-voltage coils. Save your money! If the distributor is not worn out, there is little that you can do to make the ignition system work better. Even the resistive-type cables and wide-space late-model "smog-type" rotor and cap won't cost you any horsepower if they are in good condition. The only reason to change them is if they are old and worn out. The stock coil works fine for high-performance. If yours has quit, try a Delco 1115207 coil, but don't expect any added HP. You can buy higher voltage coils from specialty ignition manufacturers, but good parts are usually cheaper at the Chevy store. If your stock one makes sparks, save your money for parts that you really need.

Chevrolet spark plug and coil cables are the "TVR" resistive noise-suppressor kind which reduce radio interference. Cut one apart and you will find that there is no wire inside of the insulation—just a piece of graphited string. You must be careful when installing and removing this type of wire because any sharp bend or yank will break the connector inside of the insulation. You won't be able to tell that this has happened unless you use an ohmmeter to measure the total resistance of the wire. As a good practice, you should replace the plug wires every two years or so—and more often in smog-ridden areas because smog really wrecks insulation.

With coil, plugs and wiring under control, the only other area of concern in the ignition system is the distributor. There is more hokus pocus and more misinformation about modifying this one component than any other part of the engine. A stock distributor will feed an engine up to 5500 RPM before exhibiting point bounce. That's an average figure—so you'll know what to expect from the stock unit. Delco's heavy-duty point set (D112PHD) will go more than 8000 RPM without bouncing. The new Chevy coils make the long saturation characteristics of dual points unnecessary. Heavy-duty points, such as those made by Accel and Mallory, allow very-high-RPM operation without point-caused problems. However, the extremely stiff springs used in these units promote rapid wear of the point rubbing block. Thus, the point setting and timing must be checked regularly or your performance will "go away" as the points close and the timing retards.

Greatly increased advance in the lower-RPM range can usually be used in cars which will be primarily engaged in acceleration contests (drags). This is because the engine passes through the low RPM range very quickly, or is seldom operated there at all. Also, long-overlap cams pump part of the fuel charge back into the manifolds at low speeds, even at full throttle, reducing charge density and allowing more advance to be used. Thus, mechanical advance-curve changes can provide startling full-throttle-acceleration improvement. A total centrifugal advance of 26 crankshaft degrees should be "all in" by 3000 RPM or even sooner for best performance. There are two basic curves for this engine that work. One is designed for street use; the other for racing. No stock-type distributor comes stock with the correct curve for either street performance or the race track, so you'll have to head for a distributor machine no matter which distributor you start with. To get the distributor to advance quicker, lighter springs can be installed. Inexpensive kits supplied by several manufacturers can be used to make this an easy task. However, to be sure that the distributor is not altered to provide more than the $12°$ to $13°$ distributor advance ($24°$ to $26°$ crankshaft) which the unit had originally, check the distributor on an ignition machine—or use your timing light on your degreed harmonic balancer to check that the recommended maximum total 38 to $42°$ BTDC advance is not exceeded with the suggested $14°$ BTDC static setting. If it is, you'll have to reduce the amount of advance in the distributor or use a less advanced static setting. The vacuum line to the distributor should be blocked off when making such checks.

The same modified curves can be applied to magnetic-pulse-type distributors because the centrifugal-advance mechanisms are the same for both point and mag-pulse types.

The full-race curve should just begin at 500 RPM (distributor); it should be smooth and gentle up to 1500 RPM with a consistent $10°$ advance showing in the distributor (20 crankshaft degrees). The curve should end at 1500 RPM (distributor). Add another $18°$ static advance (crankshaft) into the system for a total of $38°$ (on some engines this can be advanced to $40°$). With this curve there may be a

These two parts can be used to convert any rat-motor distributor into a mag-pulse triggered type, provided that the correct amplifier is also added into the system. The text explains possible combinations.

Single-point big-block distributor with rotor removed to expose centrifugal-advance weights and springs. Increasing RPM causes weights to overcome spring tension, moving outward to advance cam and thereby advancing timing.

DIESELING—Because of emission control "tuning" at the factory—the problem of "running-on" is increasing. We will limit our comment on this to the observation that engines are move likely to run-on due to a retarded spark rather than advanced spark—and that advancing spark (to a point) tends to aid engine response and acceleration. Part of the problem is caused by the curb-idle throttle-plate settings used on emission-controlled engines.

tendency for the engine to "run on" after turning off the ignition because of high cylinder-head temperatures. This, and hard starting because of static advance setting, are two of the reasons the curve is not recommended for street use. Point dwell should be 30°.

The street curve runs the same 38° total advance with 20° initial lead at the crank. This must not vary to about 600 RPM (distributor) at which point it should kick up fast to full advance of 9° (distributor) by 1,500 RPM (distributor). It is essential that the distributor return to the initial setting at idle or it will be almost impossible to get a stable idle and the engine will tend to die. This is one of the problems of using very weak springs which cause the advance curve to start kicking in close to the idle speed. Use the heavy-duty points and 30° dwell setting.

Remember that neither of these curves "creates" more horsepower or torque when the engine is smoking the tires through the quarter of a mile. The curves do increase the power at the bottom end of the RPM scale. Both curves described here are effective when used with any cam which is streetable.

Distributor Installation

Take the distributor out of the engine to work on it, period. Don't try to install special weights and springs or new points without getting that rascal out in the open where you can check it thoroughly for end play, shaft wear, smooth operation—and for actual advance characteristics by putting the distributor onto a distributor machine. For some reason, pulling a distributor just doesn't seem to be important to some mechanics or enthusiasts. Various reasons are given, including, "It takes too long," "It's not hard to work on it in the car," or, "I don't have a distributor machine." The real reason could be a fear of not being able to reinstall the distributor correctly. This is no big deal and no really special knowledge is required. Anyone can do it, including you.

While it can take a little coaxing to get the distributor gear and oil-pump drive tang to mesh sometimes . . . and it is possible to get the gear off by one tooth so that the rotor is "out of register," these little obstacles do not offset the advantages of having the distributor out of the car where you can get a good look at it without straining your eyes and your back.

Always aim the rotor in the same direction before pulling the distributor out of the engine. Either aim it at No. 1 cable in the cap, straight back or straight forward, but always do it the same and you won't have to remember what was happening. Before you do anything else, make a drawing of the distributor showing the rotor position that you selected.

The absolute easiest and surest method is to always aim the rotor at No. 1 and make sure that the timing mark on the harmonic balancer is correctly aligned for the specification static advance. With a clutch-equipped car you can push the car in high gear until you have achieved the correct conditions. It is somewhat more difficult—but not impossible— with an auto-trans-equipped car. In this instance, you can "bump" the starter to get close to the correct conditions, then use a wrench to turn the crank with the harmonic balancer attachment screw. When you have done this, reinstallation of the distributor becomes dirt-simple.

Before you take the clamp loose, scribe a line on the distributor housing and onto the block. Take the clamp off, then carefully pull the distributor out of the block, noting that the rotor turns counter-clockwise (backward) as you lift the distributor out. Note or mark the distributor edge so that you will know where the rotor turned to as the distributor finally came loose from the cam gear.

If you point the rotor in this direction and watch that scribe mark which you made, you can literally drop the distributor back into the engine—assuming that you took the car out of gear and that no one bumped the starter for you while you had the distributor out. If the distributor does not drop into place, check that the oil pump slot has not moved. A slight movement of this slot may be required so that the driving pin on the underside of the distributor gear will slip in o.k.

VACUUM ADVANCE

Don't rush to disconnect the vacuum advance which the factory included on your stock distributor. And, don't think that you should buy a replacement distributor without a vacuum advance because "real racers" use centrifugal-advance-only distributors. They don't drive their cars on the street, so the equipment which they use is not the hot tip for your street-driven car. Vacuum-advance mechanisms provide advance in relation to load and should really be called *load-compensation advance*.

Over-simplified explanations merely state that burning the fuel charge takes a certain amount of time and that burning time remains much the same regardless of RPM. Because the compression stroke requires less time as RPM increases, ignition must start earlier to allow burning to be completed at an ideal time in the stroke. T'aint *necessarily* so! Such simple explanations completely ignore combustion-chamber turbulence which hastens burning. Turbulence moves unburned mixture into the area where ignition is occurring—instead of relying on slow travel of the combustion flame front. Moving the unburned mixture into the combustion area speeds up burning so that there's usually no necessity for the advance mechanism to continue operating beyond 3000 RPM or so.

Charge density further complicates the picture. A light charge—as created by part-throttle operation—is a "slow" burner . . . needing earlier ignition. Up to 20° additional advance can be tolerated by the big block under certain part-throttle conditions. And, we may see this amount increase still further as the emission-reducing trend away from "quench-type" combustion chambers continues. However, the hotter versions are usually restricted to about 12° (crankshaft) vacuum advance to avoid the rattle which would occur when the throttle is opened quickly.

A normal appendage on all big-block Chevy distributors (except one) is a vacuum-operated diaphragm. At light throttle, high manifold vacuum advances the spark a considerable amount. As the throttle is opened further for acceleration or to climb a hill, etc., the manifold vacuum drops off and the vacuum-advance diaphragm relaxes so that the distributor assumes a more-retarded position. The vacuum diaphragm

Solid or stranded metal wire is often touted as the only way to get full power to the spark plugs. However, such wire radiates energy in the radio/TV frequency spectrum, causing unwanted and illegal interference. Magnetic-suppression spark-plug wire meets FCC requirements for high-frequency suppression without causing any loss in engine performance. Monel metal wire wound around a magnetic core stops unwanted radiation; gives negligible DC resistance. This is the only type of interference-suppression wire which can be used with capacitive-discharge ignition systems. Enlarged view illustrates MSW construction.

mechanism supplements the centrifugal-advance mechanism—which provides RPM-compensated spark advance.

Take off the vacuum advance—or disable it by disconnecting the hose—and you can expect awful gas mileage and spark-plug fouling because the engine is forced to operate with the spark retarded under cruising conditions. This reduces thermal efficiency of the engine. The spark plugs foul because they cool off, allowing soot to form on the porcelains and misfiring immediately follows because the spark travels along the soot to ground ineffectively—instead of jumping the electrodes to fire the mixture.

If you have installed a long-duration cam which reduces the manifold vacuum at low speeds—or installed a larger carburetor which drops the manifold vacuum more quickly on quick throttle opening—you may want to alter the vacuum advance mechanism. According to Jere Stahl, the spring tension against which the vacuum must work to advance the spark—and the amount of vacuum-caused advance—can both be reduced to make the car more drivable for street use. He suggests limiting diaphragm travel to 8-10° (crank) and changing the spring tension so that the diaphragm will back off more quickly. He claims that this greatly aids economy on hotrodded big blocks which are driven on the street. Using such "tricks" reduces the rattle or hard pinging that is such a familiar sound when the throttles of a high-performance engine are suddenly opened during street driving.

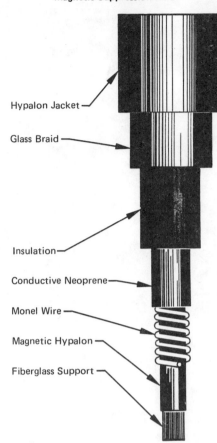

Magnetic Suppression Wire
- Hypalon Jacket
- Glass Braid
- Insulation
- Conductive Neoprene
- Monel Wire
- Magnetic Hypalon
- Fiberglass Support

Single Holley 4500 carburetes this big-block on Arlen Kurtis' record-holding drag ski boat. Manifold is a Weiand Hi-Ram.

SPECIAL IGNITIONS

Glamorous transistor or capacitive-discharge ignition systems are interesting technical achievements, *but they are expensive!* Their real advantages include long point life, easier starting in cold or damp weather and increased spark-plug life. This latter feature is the real "plus" of the capacitive-discharge systems because they will fire plugs which are too worn out for peak performance with a stock ignition system. But, special ignition systems *do not increase acceleration or top speed and they do not give more gasoline mileage than the stock ignition system in good condition.* Peak ignition requirements, according to Champion engineers, are for maximum voltage at the plugs when the engine is being accelerated at low speeds under full throttle. Transistor and capacitive-discharge systems are equal to — *but not better than* — stock systems under these conditions.

Buy a capacitive-discharge unit if you are choosing between these two types.

And, if you can afford it, buy the magnetic-pulse-triggered-type distributor with matching amplifier or converter. With a magnetically triggered system there are no points and no dwell settings. This eliminates all future worries about points wearing out the rubbing block so that dwell, coil-saturation time and advance are reduced. These systems also eliminate the need for frequent inspections of points so that this time can be devoted to other maintenance. Actually, this is the most troublefree type ignition system you can install on your big block. That's why many CanAm racers and professional drag and circle-track racers figure the advantages are worth the price. We'll probably see more and more stock car cars equipped with these systems as the costs for automotive electronics continue to be reduced — and as the already-too-high costs for automotive maintenance continue to soar.

A magneto ignition is the perfect answer for super-lightweight dragsters or circle-track racers which will be push-started and therefore do not need a battery. If you can really use a magneto to good advantage, then we'd recommend a Joe Hunt-adapted Scintilla Vertex one. But, if a battery must be carried to operate the starter — stick with the stock ignition and use the money you save to buy some helpful item of equipment which will add performance. The only time that a special ignition will produce more HP than the stock coil and distributor is when the stock system has a problem or the plugs are worn out.

Several special systems are pictured in this chapter with captions explaining the highlights of each.

SPARK PLUGS

Don't be surprised if the spark plugs "go away" in a hurry. The big Chevy is noted for "eating" plugs. This is another way of saying that plug life in this engine is short!

Spark-plug life is a real problem in hotrodded engines. Even stockers can seldom get 10,000-mile plug life, even with a cleaning at 5,000 miles. The problem is complicated when you modify the engine for more horsepower and use the engine in varying ways—such as freeway driving, occasional drag races and the usual trips to work and the grocery store.

The reason for this is that combustion-chamber temperatures are considerably higher than those of most other engines. Any plugs used in the large Chevy usually fall off sharply in efficiency after only two or three thousand miles—or after 10 to 20 hard acceleration runs—especially if the stock too-lean-for-performance jetting is used in the carburetor. Some enthusiasts attempt to "cure" the problem by installing a much colder range plug for street application. This leads to premature loading up of the plugs—and thus still shorter plug life. In desperation, others have installed "guaranteed, life-time, more horsepower, mileage and engine life" trick plugs. Don't bother. No one winning any races uses them—so why should you? Most of these plugs, easily identified by the full pages used to advertise them, appeal to the "something-for-nothing" owner who thinks that screwing in new plugs will give instant and enormous improvements in HP and mileage. We know that this is impossible and so do you. But, Barnum was right, there *is* a sucker born every minute.

The words "hot" or "cold," used in reference to spark plugs, often cause confusion because a "hot" plug is normally used in a "cold" low-horsepower engine and a "cold" plug in a "hot" high-performance engine. The terms refer to the heat rating or thermal characteristics of the plug—more specifically—the plug's ability to transfer heat from the firing end into the engine cylinder head and thence into the coolant.

Three types of spark plugs are used in big-block heads. All are 14 mm, but the reach and sealing methods differ. 5/8" hex plugs with a 0.460-inch reach and a tapered seat seal against the head (no gasket) on 1970 and later engines with iron heads. Aluminum-headed Chevys get a 3/4" reach plug with threads all the way to the gasket area. Pre-1970 iron heads use 3/4-inch reach plugs which are only partially threaded—at least on AC types.

By definition, a cold-running plug transfers heat rapidly from the firing end. It is used to avoid overheating where combustion-chamber or cylinder-head temperatures are relatively high. A hot-running plug has a slower rate of heat transfer and is used to avoid fouling where combustion chamber temperature is relatively low.

The length of the core nose and electrode alloy material are the primary factors in establishing the heat range of a particular spark plug design. Hot plugs have relatively long insulator noses with long heat-transfer paths. Cold plugs have much shorter insulator-nose lengths and thus transfer heat more rapidly.

If the engine ignition timing, camshaft, compression ratio or carburetion is changed, heat-range substitution may be necessary. Stock-heat-range plugs recommended in the service manual may be replaced with *slightly colder* ones when you modify the engine. But cold plugs required for best all-out racing performance will not work for normal street/highway driving. They will make the engine spit back excessively, make gasoline mileage terrible, and the plugs will wear out or load up in a hurry, requiring frequent cleaning and/or replacement. On the other hand, a too-hot plug will last only a few miles and can cause destructive preignition. Extremely cold racing plugs do give better preignition protection *when fired by capacitor-discharge systems.* Keep in mind, though, that plugs thermally matched to combustion-chamber temperature are superior to too-cold plugs, regardless of the type of ignition system you are using.

"Hot" coils, special magnetos and transistorized or capacitor-discharge ignition systems do not require changes in spark-plug heat range. As previously mentioned, combustion-chamber temperature, not spark temperature, influences heat range. Spark erosion of the plug electrodes cannot be materially reduced by colder heat-range substitution, but erosion can be reduced with a C-D system.

A near-perfect plug is the projected core-nose which acts as a "warm" plug at low speeds and cools off to act as a "cold" plug as RPM's are increased. Some of these also have an auxiliary gap which increases plug life between replacement or regapping because it changes the spark-voltage characteristic. When the spark has jumped the auxiliary gap it has sufficient voltage so that the electrode path will be used, instead of leaking away ineffectually across fouling deposits which may have built up on the porcelain and shell.

Serious enthusiasts always run projected core-nose plugs around town and switch to colder plugs for dragging. There's just too wide a heat-range requirement for any one plug to do the entire job in a hotrodded engine, although the projected core-nose ones come close.

One of the small items which make life a little easier around the performance enthusiast's garage is a plug board. A two-foot length of 2″ x 6″ holds a lot of spark plugs—electrodes up for easy reading.

R. J. Gail, Champion Spark Plug's Racing Coordinator says, "The projected core-nose plug benefits from the cooling effect of the incoming charge at high engine speeds. This 'charge cooling' on the projected firing end provides greater preignition protection at high speed and makes it possible to increase the insulator length on a given design, thereby improving the spark-plug-temperature characteristics. In many racing applications, with critical spark-advance settings, the possibility of core-nose fracture from detonation would be more likely with projected core-nose plugs than with regular types."

The stock plug gap of 0.035-inch is fine for a modified engine. Setting extra-wide plug gaps because of exotic ignition systems presents risks. If ignition becomes marginal in a race, a closer gap might prevent the breakdown.

Plug life and consequently, replacement expense, can be enough of a problem to make buying a capacitive-discharge ignition system very worthwhile. Some Porsche 911's, for instance, are factory-equipped with capacitive-discharge ignition systems to overcome their short plug life.

According to an article in CAR LIFE Magazine, July 1968, a capacitive-discharge system increases usable spark plug life two to five times. Such systems greatly reduce the ills which may result from inadequate ignition when the conventional system nears tuneup time.

Capacitive-discharge ignition systems have the ability to fire fouled plugs and even worn-out plugs ... as if they were new ones. Engines with C-D ignition systems can perform o.k. with plugs which are one to two heat ranges colder than stock, providing performance similar to that which you'd get with new projected core-nose plugs. If you are running a rev-limiting C-D ignition such as a Howard Speed-A-Tron or an ARE unit, colder plugs are necessary to keep "glo-plugging" from cancelling out the speed-limiting effects.

The reach or thread length of the plug should be correct for the type of head you are using. Pre-71 iron heads use a very long reach plug with a 13/16-inch hex. All of the aluminum heads through 1971 use an extended reach "XL" plug with 13/16-inch hex. All 1971-and-later iron heads use a taper-seat 5/8-inch hex plug with short reach. It is very important to check plug-thread length when you have the cylinder heads off the engine because exposed threads on the spark plug—or in the head itself—get red hot and cause destructive detonation that can burn or "hole" pistons.

When substituting a plug with a slightly longer reach, as you might have to do if plugs of the correct heat range were not available in the required reach, a thicker plug washer must be used so that no threads will be exposed in the chamber. *Do not use multiple washers of the crimped-copper or crimped-aluminum type.* They will not conduct the heat out of the plug correctly. Champion sells standard, 0.060, 0.080 and 0.100-inch thick washers as stock items. Or, you may find a copper oil-drain-plug washer of the correct thickness for this application. Plugs with a shorter reach can never be used under any circumstances because these would leave the sharp threads in the head exposed to the combustion process.

Do not be quick to condemn a spark plug if it does not spark in a plug tester at an air pressure equal to or greater than the engine cranking pressure. This idea is entirely wrong because air pressure as read on a plug tester has no direct relationship to engine-cranking pressure for several reasons. In a plug tester, the plug stays cold. In the engine, the electrodes operate at high temperature and require less voltage to fire than when cold. Keep in mind that spark plugs normally do not fire at the point of maximum cylinder pressure, but well ahead of top dead center. For these and a number of other reasons, the efficiency of a spark plug cannot be measured in terms of "pounds per square inch"—which is what the plug tester attempts to establish.

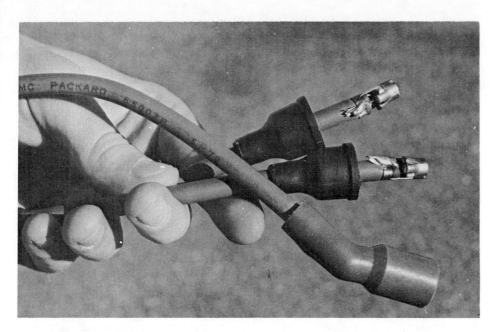

Chevy has these silicone-compound Packard 65003R wires for you. The complete soldered-terminal, steel-core set is listed in the HD Parts List. The insulation is heat resistant so that the cables will not be destroyed if you happen to lay one against a hot header someday while you are changing plugs.

Spark Plug Cables

Cross-firing or induction leakage between cables will occur when ignition cables are grouped closely together and run in parallel for some distance. This causes engine roughness and can result in damaging preignition and detonation.

Cross-fire is not caused by defective cable alone. It is traceable to the magnetic field which surrounds any high-tension conductor. Thus the lead which is carrying high voltage at any given instant tends to induce voltage into an adjacent lead. Troublesome cross fire is most likely to occur between consecutive firing cylinders when these cylinders are located closely together in the engine block. Never tape spark plug cables together and don't run spark plug wires through metal looms—no matter how neat they look!

Use the stock wire separators. Keep the cables in the correct slots and route the wires per factory recommendations so that cross firing cannot occur.

For racing, where plugs are often inspected and/or changed, install the silicone rubber-insulated wire with stranded-steel wires soldered, not just crimped, to connectors at each end. This sparkplug-wire set is in the HD parts list, page 150. Silicone insulation is especially good where plug wires must be routed near exhaust headers, or where the wires might be laid against the exhaust headers as you change plugs. It's the wire of choice for street-driven cars with hot-running emission-controlled engines, too.

Stock TVR cables do not work well with capacitive-discharge ignition systems. In fact, they can be expected to fail completely. Ted Trevor of Crown Manufacturing suggests using "MSW" (Magnetic-Suppression Wire) which is a coiled steel wire in spark-plug cable insulation. This type of wire will not fail when used with C-D systems and it suppresses ignition noise radiation so that you can listen to the radio. The use of the stranded-steel ignition cables typically used for high performance will destroy radio listenability for you and anyone who happens to be driving near you.

How to "read" plugs

Selection of the correct carburetor or injector jets can be aided by "reading" the appearance of the plug electrodes and porcelain insulator shell. The accompanying plug color chart explains what to look for. Because the part of most interest is the base of the porcelain—which is "buried" in the plug shell, a magnifier-type illuminated viewer as made by AC or Champion should be an early purchase for your racing tool box.

Remember that new plugs will take time to "color." As many as three to four dragstrip runs may be required to "color" new plugs. If you are drag racing with plugs that have been run on the street—and we can't imagine why you'd want to—then keep in mind the fact that the plugs *may* clean out after several hard runs down the strip, but they will probably speckle the porcelain in the process.

Plug color is only meaningful when the engine is declutched and the engine "cut" clean at the end of a high-speed full-throttle, high-gear run. If you allow the car to slow with the engine still running, plug appearance will be meaningless. Plug readings can be made after full-throttle runs on a chassis dyno with the transmission in an intermediate gear so that the dyno is not overspeeded—but road tests require the use of high gear to load the engine correctly. Similarly, plug checks can be made where the engine has been running at full throttle against full load as applied by an engine dyno. It is easier to get good plug readings on the dyno because full power can be applied and the engine cut clean. Plugs can be read very quickly because you can get to them easier—as opposed to the usual car or boat installation. However, don't think that plug heat range and carburetion jetting established on an engine dyno will be absolutely right for the same engine installed in your racing chassis or boat. Conditions of air flow past the carburetor or injectors can easily change the requirements —and perhaps unevenly so that different cylinders need different changes.

It would be nice if every plug removed from a racing engine should look like the others from the same engine—in color and condition—but this is not easily achieved. Differences in color or condition indicate that combustion-chamber temperatures or fuel/air ratios are not the same in every cylinder, or that related engine components need attention. The problem is greatly complicated in the large-block Chevrolet by the differences in turbulence and efficiency of the cylinders. The cylinders with intake ports which bring the fuel into the center of the cylinder operate differently from those with ports which direct the charge into one side of the cylinder against the wall. Thus, you should expect plugs used in cylinders 1, 4, 5 and 8 to look similar. Cylinders 2, 3, 6 and 7 should also look like one another.

If differences exist in the condition of the firing end of the plugs, the cause may be found in one or a combination of factors: unequal valve timing, weak compression, poor oil control (rings), weak ignition system, unequal cooling around the plugs or unequal distribution of the fuel/air mixture.

If you correctly followed the procedures for blueprinting the engine then you should obtain equal cylinder pressure *at cranking speeds*. However, unequal cylinder pressures can lead to one or more plugs running "off." Check for improper valve lash, defective or unseated rings, a blown head gasket or severe differences in clearance on the cylinder walls. Problems within the ignition system which might lead to plugs not reading the same or misfiring include: a loose point plate, arcing within the distributor cap, a defective rotor or cap, cross-fire along with primary wiring, or a resistor which is opening intermittently. Pay special attention to cleanliness of the ignition system, including the inside and outside of the distributor cap and the outside of the coil "tower." Also clean the inside of the cable receptacles on the coil and in the cap. Any dirt or grease here can allow some or all of the spark energy to leak away.

You'll need two sets

Best performance in "strictly stock" drag racing or high-speed road work is not usually obtained by using the same plugs for street and race driving. Street plugs pick up deposits during low-speed running . . . deposits which tend to bleed away voltage (lessening the spark intensity) under high crankshaft speeds and heavy loadings. These deposits do not fluff off the plugs, but melt to form a conductive coating which causes misfiring. Thus, for a double-duty engine *at least* two sets of plugs are required: one set with heat range ideal for the street and another with the heat range required for the strip. Race plugs should be gapped closer—depending on the amount of compression and grade of fuel. Although a closer gap makes for an erratic idle, who worries about that when it's time to go racing?

A capacitive-discharge system creates a very "fat" spark which allows using plug gaps 0.002" to 0.005" wider than with the stock system.

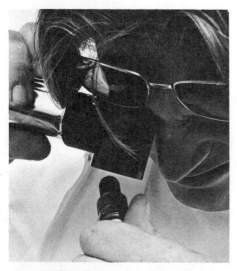

Gary Loftin checks color and appearance of plug porcelain on one of the sparkers from the ZL-1 engine in his drag boat. One of these Champion Sparkplug Viewers, No. CT456, should be in every big-block tuner's tool kit.

PLUG COLOR CHART

Rich - Sooty or wet plug bases, dark exhaust valves.

Correct - Light-brown color on porcelains, exhaust valves red-brown clay color. Plug base slightly sooty (leaves slight soot mark on hand when plug base turned against palm). New plugs start to color at the base of the porcelain and this is only visible with an illuminated plug magnifier.

Lean - Plug base ash grey. Glazed-brown appearance of porcelain may also indicate too-hot plugs. Exhaust valves whitish color.

NOTE: Piston top color, as observed with an inspection light, can be a quicker and sometimes more positive indicator of mixture than plug porcelain appearance. The careful tuner will look at all of the indicators to take advantage of every possible clue as to how the engine is working. Inspection lights are available from Moon Equipment.

Tom Jacobson's big-block-powered Chevelle has Weiand Hi-Ram manifold with two Holley vacuum-secondary four barrels. Although all four of the side-hung bowls are plumbed to the fuel "log" — this is not needed. Note fuel-pressure gage and in-line fuel filter.

DETONATION AND PRE-IGNITION

Detonation or "ping" is a sharp knock which makes your engine ring or sing as if someone had hit it with a hammer. Skip Mason of Champion Plugs says, "It is like hitting the pistons, valves, spark plugs and other engine parts with a hammer." There is usually no problem in identifying detonation because you can hear it—unless you are running the engine with open exhaust. *Detonation is spontaneous combustion of fuel in the chamber instead of the desired controlled "slow" and even burning.* Perhaps the most common cause of detonation is over-advanced ignition timing, but there are other causes which should not be overlooked. It can also be caused by full-throttle acceleration with too-high a compression ratio; high operating temperatures, too-hot spark plugs, and lean mixtures can all be contributing factors to cause destructive detonation. Detonation results include hammered-out bearings, broken rods and crankshafts, sharp-edged holes in piston crowns, and broken ground electrodes on spark plugs. It also breaks ring lands and cracks piston skirts.

Pre-ignition is ignition of the fuel while the compression stroke is occurring, but much earlier than intended. It can be caused by glowing plug, sharp valve edges, overhanging gaskets. Preignition can also be caused by detergent oil when barium and calcium deposits get hot. If caused by a too-hot plug, the first clue will usually be the erosion of the center or ground electrode. In this case, a colder plug may cure the problem. A tuliped intake valve is *always* a sign of preignition. This can show up when the problem was not severe enough to scuff a bore or hole a piston. Preignition can be temporarily reduced by retarding the spark, but this creates excessive heat and detonation sets in due to this heat—as a result the fuel fires as it enters the chamber. Complete chaos and a destroyed engine inevitably result from pre-ignition.

DO NOT DYNO-SET IGNITION TIMING

Maximum HP dyno-set ignition timing will usually be too far advanced for *any* type of driving. Such settings cause detonation because spark-chamber turbulence, and because spark-plug temperature is greatly increased by advancing the spark setting. Detonation results in either case and can also lead to pre-ignition in the worst cases. This all happens because the chassis-dyno operator sets the distributor for the best *flash* reading, and you drive off the rollers with the engine detonating merrily. Worse yet, you may not drive the car after setting the timing on the dyno. If you enter the car in competition the chances are very good that you will not hear the engine detonating because exhaust noise will mask the detonation. The dyno operator is not likely to pay you for a blown-up engine. It is essential that you understand what can happen to your engine.

While it is possible to cheat a little on the factory settings by advancing a few more degrees beyond the shop-manual recommendations, detonation must be listened for and the spark retarded if it occurs. If you hear detonation and the engine is not being run with close to 38° total advance (crankshaft degrees), try increasing the size of the main jet. Or, look for exposed plug threads or a sharp edge on a valve or somewhere else in the combustion chamber. Excessive carbon build-up can also be the culprit.

Barney Navarro, in a 1963 POPULAR HOTRODDING article, made the following observations. "Among the many things that affect the spark-advance requirements of an engine, we find engine temperature and air temperature. The hotter the fuel charge before ignition, the faster it burns, therefore requiring less spark advance. The speed of any chemical reaction is doubled by a temperature increase of 18° Fahrenheit, so it is easy to see that distributors should be fitted with temperature-compensating devices, especially those of air-cooled engines. The average modern overhead-valve engine can utilize from 5 to 8 degrees more advance when it is cold than when it reaches operating temperature."

Navarro's comments tell us why we can get away with cranking up the advance for quick blasts on the dyno or a fast trip down the quarter mile—but destroy the engine by trying to use the over-advanced distributor setting for everyday driving.

High-perf Chevys that do most of their running above 6,000 RPM sometimes close up the spark-plug electrodes. Each plug should be marked with a line on the insulator to indicate the "bend" side of the electrode. The line can be drawn with a marking pen, fingernail polish or paint. Keep the line upward when installing the plug, even if you have to switch plug washers to make this happen. Positioning the electrode in this fashion may even aid flame travel in the chamber.

Rev limiters and tachometers: page 124

Torque specifications
twist values keep 'em running strong

Size	Use	Torque (lb. ft. unless noted)	Lubricant[1] (oil, unless noted)
1/4-20	Oil-Pump Cover	80 lb. in.	
	Oil Pan to Front Cover	80 lb. in.	
	Front-Cover Bolt	75 lb. in.	
	Rocker Cover	25 lb. in.	Antiseize
5/16-18	Oil-Pan Bolt	165 lb. in.	
	Camshaft Sprocket	20	
3/8-16	Intake Manifold	25	
	Water Outlet	20	
	Exhaust Manifold	20	Antiseize
	Clutch Pressure Plate	35	
	Distributor Clamp	20	
	Flywheel Housing	30	
	Water Pump	30	Sealant
3/8-24	Con-Rod Bolt	50^2 (0.0055-in. stretch)	
7/16-14	Cylinder-Head Bolt	Long 75	Sealant
		Short 65	Sealant
	Cylinder-Head Stud	Long 65	Sealant
		Short 55	Sealant
	Valley Stud (alum. head)	50-55	Sealant
	Rocker-Arm Stud	50	
	Oil Pump	65	
7/16-20	Con-Rod Bolt 3959184	$60\text{-}65^2$ (0.007-in. stretch)	
	Con-Rod Bolt 3969864	$67\text{-}73^2$ (0.009-in. stretch)	
1/2-13	Main-Bearing Bolts Inner & Outer	110 (iron block) 100 (aluminum block)	Molykote
1/2-20	Harmonic Balancer	85	
	Oil Filter	25	
	Oil Drain Plug	20	
1/2-14	Temperature Sender	20	
14mm	Spark Plugs 13/16" Hex	25 (iron heads) 25 (aluminum heads)	Antiseize
	Spark Plugs 5/8" Hex	15	

TORQUE SEQUENCE

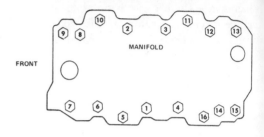

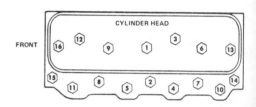

NOTES:
1. Many racing engine builders use Loctite instead of oil to eliminate possible problems caused by bolts or nuts loosening.
2. Although a rod-bolt torque "range" is recommended by Chevrolet, engine builders have found that the 10B39 boron-steel rod bolts often require 80 to 110 lbs. ft. torque to get the recommended 0.009-inch stretch. This stretch is essential to preload the bolt so that the nut will not come loose as the rod bolt head and nut "skin" into the rod body and cap during operation. Stretch is the preferred specification, even if torque exceeds values shown. It is a good idea to record the torque required to obtain the stretch for each bolt. Then if you have to check the lower end and get it back together in a hurry someday, you can torque each bolt to your prerecorded specifications without having to back the nuts off and retorque.

Camshaft & valve train
understand what's happening before you change it

Because the camshaft is seemingly easy to remove and reinstall in the Chevy, it is often one of the first items which the novice changes when chasing after additional power. But, it needs very careful installation to ensure that maximum benefit is obtained from its action and the knowledgeable tuner changes the cam only after he's "made" all the HP he could with the stock cam. And, then he may decide to use another stock Chevrolet cam.

The camshaft is only part of the valve train and the other items deserve equal attention if utmost performance is to be achieved. Lifters, pushrods, rocker arms, spring retainers, and valve springs interact in ways which are not completely understood by many mechanics. Installation techniques are also important because they affect camshaft life. The coverage devoted to these related parts and to careful installation will seem over-long to the casual reader, but the thoughtful engine builder will recognize that a short presentation does not give the complete picture.

Several camshafts have been offered in stock engines and each cam is well mated to its stock engine. Grinders of special cams can help with cams specifically designed to meet the requirements of non-stock applications. A detailed discussion of effective valve opening, as well as timing graphs relating stock cams to reground ones, will dispel some of the hocus-pocus surrounding the entire subject. Valve spring and retainer information is also included with specific directions for checking and changing camshaft timing at installation. Rocker-arm geometry and its effect on valve lift and guide wear is fully covered. The reader who follows all of this information with great care will discover that his engine has suddenly gained "free" horsepower—even with a stock cam—and that is the best kind!

HOW THE CAMSHAFT WORKS

It looks like a simple stick with bumps on it, but the cam is the major controlling element in determining your engine's ultimate behavior. The stock camshaft is carefully related to a fantastic number of factors, including carburetor area, compression, displacement, transmission and axle ratios, car weight, performance desires, driving habits of the buying public, and so on—ad infinitum. You cannot relate these factors to the same fine degree as the factory engineers with their experience and computers. We can consider what goes on during the intake and exhaust operations in a single cylinder.

Valve timing is related to degrees of crankshaft rotation, but avoid this oversimplification. Another relationship is more helpful in describing camshaft functioning: valve action versus displacement of the cylinder. Where is the piston and what is it doing?

As high-school students we were taught four-stroke internal-combustion-engine functioning in four unrelated 180-degree segments: Intake, Compression, Power, and Exhaust. If the valve mechanism could slam the valves open and shut precisely at the TDC (Top Dead Center) and BDC (Bottom Dead Center) points without delays, the engine would probably run—until some part broke! But, just as instantaneous valve action is impossible due to the severe forces such would impose on the valve mechanism, instantaneous movement of the intake and exhaust gases is impossible. So, the 180°-segmented description of valve opening versus crankshaft position does not work outside of the textbook covers. Now that another school-boy belief has been destroyed, let's look at practical facts.

Top side of Bill Jenkins' Pro Stock engine. Those large-diameter pushrods are heavy-duty Chevy—listed in our compilation of heavy-duty parts.

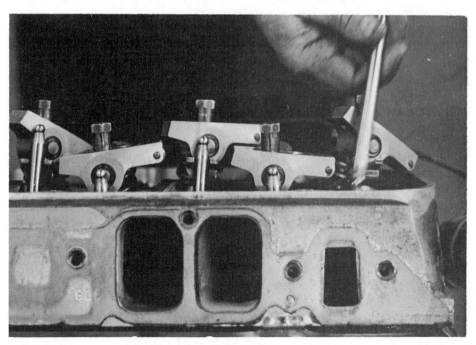

The intake valve is opened *before* TDC while the piston is still rising on the exhaust stroke. There is not much pressure left in the cylinder and the early opening is added to the textbook's 180°, even on "mild" engines. Valve acceleration rate can't exceed certain limits, so the cam designer opens the valve fully before the piston reaches its point of maximum acceleration. Otherwise, a partially open valve would impede gas flow just as cylinder displacement is being increased at the most-rapid rate. That's exactly what happens with the low-lift hydraulic cam and to a limited degree with the high-lift hydraulic cam. Opening the intake at almost TDC gives economy *and* low-speed torque beyond what you'd ordinarily get. Early opening is often touted as a cleaning device for pushing the exhaust out and helping to clean up the chamber of spent gases. According to Racer Brown, this may be true of a good hemispherical chamber with a tuned exhaust and intake system, but it certainly does not apply to both big-block chamber types. Even at best, this so-called "scavenging action" is incomplete. Historically, there have been a great many more bad combustion-chamber designs, hemispherical or otherwise, then there have been good ones. Early opening in the big block should be approached with great care because the incoming gas has a decided tendency to do an "about-face escape act" through the convenient exhaust-valve opening. All you get for your efforts is increased fuel consumption, plus a few other disadvantages, and perhaps a bit of exhaust-valve cooling by the escaping fresh mixture.

Very early opening confuses the intake system at low speeds, causing the rough-idle trademark of a high-RPM engine. While early opening causes no particular problems at high speeds which provide an atmospheric-pressure operating condition, reduced speeds cause manifold vacuum to suck exhaust gases into the intake system. Results are diluted air/fuel mixtures, rough idling, spark-plug fouling in extreme cases, and fogging of the mixture out of the carburetor or injector throats. This dilution is one of the reasons why an accelerating racing engine tolerates full ignition advance at very low RPM.

There is good reason to leave the valve open *after* BDC to take advantage of the incoming-gas momentum. Although the piston starts its upward stroke after BDC, it is barely moving upward for quite a few degrees. The crank swings through a considerable arc to swing the connecting rod big end sideways. The momentum of the inflowing gases gives additional filling during this period of piston "laziness." As piston-acceleration rate increases, the intake valve must be closed or severe reverse pumping and charge-density reduction will occur. It should be obvious that as engine RPM is increased, the closing point can be left until later than is desirable on an engine which has to produce good torque in the low and mid range. A tuned-intake system may increase charging-mixture momentum to improve cylinder filling in a limited RPM range. Reverse-pumping of the charge back into the manifold, as caused by late closing, is one factor allowing engines to be set up for drag-race competition with much higher compression than would ordinarily be possible with gasoline for fuel. The same cars would require lower compression if set up for continuous high-RPM operation, unless running at high altitude as at Bonneville.

If we consider four points only—intake opening and closing and exhaust opening and closing, *the one point that has the most dramatic effect on power output is the point at which the intake closes.* There is a fine balancing act going on here; trying to take advantage of the inertia of the air/fuel charge, plus any late arrivals in the form of positive sonic pulses and closing the intake valve *before* these forces are overcome by reverse-pumping action.

Exhaust-valve operation can almost be deduced from what you have read about the intake, only in reverse. During the induction period, we do everything we can to fill a cylinder that is at or very near ambient atmospheric pressure. During the exhaust period, we try to unload the same cylinder of all gases, including residuals. The average pressure is six or more times that seen during the induction period. This is easier, therefore the exhaust opening and closing points are not so critical as those on the intake side.

Because the exhaust valve needs its "head start" on the piston, part of the power stroke is "subtracted" by opening the valve *before* BDC. This releases pressure from the cylinder so the piston works less to pump out the gases, and ensures that the valve is fully open by the time the piston is accelerating at its maximum upward rate. Pressure "blow-down" also starts the exhaust-gas flow out of the cylinder. Little power is lost by early exhaust-valve opening because the burning gases impart most of their available effort in the first 90° of the power stroke. For this reason some stock big-block cams are ground with dual patterns to give more exhaust than intake duration. When closing the exhaust valve *after* TDC—the intake action is overlapped—both valves are open at the same time. If this overlap is stretched too far, serious dilution occurs, especially past 70° overlap. When a tuned-exhaust system is used, late closing of the exhaust takes advantage of the momentum of the leaving gases to help discharge the cylinder. It is sometimes possible to reduce cylinder pressure on a momentary basis to below atmospheric (vacuum) so the incoming gases are started in by the draft of the exhaust gases, as well as by the increasing displacement afforded by the descending piston.

While the foregoing can be interesting in understanding the operation of your engine, it would not enable you to tell the cam grinder the precise number of degrees of overlap and duration which you would need for specific applications. Fortunately for most of us, the cam grinders use their superior knowledge of mechanical limitations and the physics of gas flow to protect us from ourselves. Cam patterns are available for every situation. You cannot go far wrong by relying on the recommendations of qualified cam grinders, especially if you are willing to follow through with careful installation procedures as will be described.

STOCK VALVE TRAIN

Chevy's valve train has been widely copied by the designers of other engines. A chain-driven camshaft of cast-iron alloy with 16 lobes has five bearing journals which operate in bearings pressed into the block. Lifters operate in holes bored in the block. Tubular-steel pushrods connect with stamped-steel rocker arms in the "tub"-shaped section atop each head. Each rocker arm pivots against a ball which locates the combination on a stud. A guide for each cylinder guides the pushrods close to their tops. Camshaft motion is multiplied by 1.7 by the rocker arm which contacts the tip of an intake or an exhaust valve.

Valve stems are 3/8-inch diameter. Seating pressure for each valve is provided by one or more coil springs. Springs are retained by a stamped- or machined-steel retainer wedged to the valve stem by valve keys. Chevrolet's steel retainers are stout designs which are suitable for racing. They weigh slightly more than aluminum ones but their extra strength, fatigue resistance and low cost make these the parts to use. A stamped-steel cover forms the outer half of the rocker-arm enclosure. Because these covers are always internally bathed in a thin film of oil, they can significantly aid in cooling the engine if you use non-chromed covers painted flat black.

STOCK CAMSHAFTS

The accompanying table indicates model, intake opening and closing, exhaust opening and closing, lift and duration for the six stock big-block cams which were available as of 1971.

You will notice that these specifications do not agree with factory published specs. This is because factory cam-timing specs are "out to lunch." There's no way that you can make heads or tails out of them because each one must have been measured differently. We measured the cams similar to the way that one cam grinder measures his—Racer Brown. Other grinders use different criteria. Because there's no single standard, a user cannot actually compare camshafts to get a valid understanding of what's what. Perhaps this chart of stock timing and lift, plus the graph showing actual valve-lift patterns for six stock big-block cams will help to sort out the hocus-pocus which surrounds the whole mysterious area of cam contours. Needless to say, the stock Chevrolet offerings can provide impressive performances when teamed correctly with carburetion, headers, compression, gear ratio, tires and clutch combinations that "get it all together."

As is the case with any production-made part, camshafts occasionally vary slightly due to the production tolerances which must be allowed to keep prices where ordinary folks like you and us'ns can afford to buy them. But you should become aware of the fact that automation has drastically improved production tolerances over the years so that factory parts are typically the most accurate that you can buy. You can disbelieve much of the scare propaganda passed out by hotrod parts makers who would have you believe that they are the original inventors of the wheel. The ones who scream loudest about quality hope that you will never measure one of their cams for timing or lift consistency or for consistency between the base circles.

STOCK CAM COMPARISON

CAM AND USE	TIMING AND LIFT[1] Intake/Exhaust	DURATION[1] Intake/Exhaust
Low-lift hydraulic 3874872 L-35, L-66, LS-1 and LS-3	10-54/79-25 0.398/0.430	244/284
High-lift hydraulic 3883986 L-34, L-36, L-68 and LS-5	20-68/74-20 0.461/0.480	268/274
Hi-perf. street & marine hydraulic 6272989	56-120/107-65 0.500/0.500	356/352
Street mechanical 3863143 L-71, L-72, L-78 and LS-6	40-86/88-38 0.500/0.500	306/306
Mechanical "off-road" 3925535 L-88	48-98/96-58 0.540/0.560	326/334
Mechanical[2] "off-road" 3959180 ZL-1	50-92/94-60 0.560/0.600	322/334
Mechanical[2] fuel-injection Can Am, "off-road" 3994094	72-114/104-76 0.600/0.600	366/360

NOTES
1. All timing and degree indications are in crankshaft degrees. Lifts are in inches. Duration measurements were made at the lifters with the hydraulic lifter at 0.008-inch lift off the base circle and the mechanical lifters at 0.015-inch lift off the base circle. With mechanical lifter cams, this is roughly equivalent to valve lift-off (of the seat) when lash is taken up. This accounts for the long durations shown.
2. 100% Magnaglo inspected at the factory with no indications permissable.

55

Most magazine articles tell you to ashcan the stock camshaft: more baloney designed to please the cam grinders who spend big $$$ advertising their products. The writers of the articles have no choice. If they want to keep their jobs they have to push the products of the manufacturers who can afford to spend money for advertising. If stock cams are so bad, howcum so many million Chevy's manage to get their owners to where they are going every day? Chevy has six stock cams and one of them may be just right if you select one that is correct for *your* application. And, the price is right!

When building up any modern high-performance engine it is difficult to avoid the temptation to "put a cam in it." In great part this is perhaps due to the penetration of advertising by those who sell camshafts for a living. Many cam grinders turn out excellent products — but they are for specific purposes. Despite advertising claims, the purpose is not usually street or highway driving. So, unless you are an old and experienced hand in the game of horsepower seeking, you would do well to consider the mechanical-lifter camshaft which has been supplied in the big-block high-performance engines from 1965-71, the 3863143. It has 0.500-inch lift for both intake and exhaust and a claimed duration of 306° for the intake and 306° for the exhaust — which will be considerably shorter on an effective-duration basis. Or, you may want to consider keeping the hydraulic lifters in your engine to ensure quiet valve-train operation. A high-lift, short-duration hydraulic-lifter cam can often provide plenty of performance for the occasional stoplight Grand Prix or a trip to the drag strip on grudge night.

Although it is easy to buy the L-88 and ZL-1 cams from your dealer, the truth of the matter is that these cams do not work well if the engine has to exhaust through mufflers. In fact, the 3863143 cam previously mentioned is superior when mufflers are required.

When you start cranking in extra duration, the engine begins to get "soggy bottom." You have to scream it up to get enough torque to move off the line or away from a stop sign—then the torque comes in "like gangbusters," usually when you don't really need or want it. That kind of engine behavior is strictly all wrong for the street, so think twice—maybe even three times—before stuffing a wild lumpstick into your go-to-work car or grocery-getter. Buy a decal instead. Decals are cheaper than cam kits anyway. If it makes you feel better, tell your friends about your "wild" camshaft and drive on by with your car in high gear and your engine going "pocketa-pocketa-pocketa." Happiness is often a stock cam.

If you are going racing or want exceptional performance for the street you may want to change the camshaft. But here again, don't be misled by the decals on your hero's car. Chances are that he's paid to run the decal, regardless of whose cam he selects. And, the decal is often changed at run-off time because *"money talks"*.

Look at the flow-data charts in the cylinder-head chapter and you'll note that the stock heads, even with 1.72 exhausts and 2.06 intakes, just keep breathing easier and easier as the lift is increased. The big-block engine "likes" a lot of lift and will nearly always run better when a cam with more lift is installed. If the duration is kept at a reasonable figure, many of the bad traits of racing cams do not show up.

Stock cams are priced *right*. The racer's "friendly price" for stock cams at most dealers will be less than that for "off road" cams. Price alone could be a good reason to use the factory parts, but bear in mind that the street-mechanical cam is as far as you should go for a street-driven machine with mufflers. The off-road cams are for open exhausts and lots of RPM, so don't be confused. If you need more HP than the street-mechanical cam offers, consider a high-lift short-duration hydraulic camshaft available from aftermarket sources.

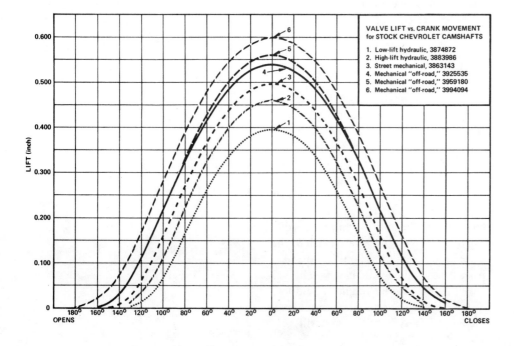

These curves show actual intake-valve lift measured in crankshaft degrees on each side of the cam lobe centerline.

RACING CAMSHAFTS

Camshaft selection from amongst the thousands of "bump sticks" which are offered can be mind boggling—unless you know and understand what you are looking for. Although there are dozens—maybe even a hundred—cam manufacturers offering big-block cams, we called on W. G. "Racer" Brown to see what might be used to wake up the engine —beyond the factory's offerings.

In fact that's where we started our questioning, "If the factory offers two hydraulic cams and four mechanical cams ranging from streetable to wild—what more can cam makers offer to the big-block owner?" Racer rose to the bait like a hungry trout. "Factory offerings have to be more than somewhat reliable, so their designs have to incorporate a more decided degree of conservatism than those which we can make and sell. Our dealers are not forced to service the entire automobile and hold still while a customer complains about a rough idle and so forth. Chevy cams are designed for either street use or far-out long-track competitive events such as Daytona for sports-car racing in Corvettes or Can Am racing in all-out fuel-injected engines. As a result there's lots of middle ground. You can't put an L-88 cam in a street machine and be happy and it won't work on a 1/4- or 1/2-mile track, either. So, that's where we come in—with duration, valve action and lift in enough different formats to allow tailoring the cam to give driveability with punch in the areas where it's needed. Because we can be bolder in our approach, we can nearly always add torque or HP in large amounts as compared to the factory cams."

Next we asked why there are not more hydraulic-lifter cams being used for fast street/strip machines. We were nearly overcome by the torrent of information here because Racer seriously thinks that hydraulic cams are the most under-rated and overlooked components around the racing scene. "Lots of people distrust hydraulic lifters because the first ones were awful. And, most folks think that hydraulics pump up to cause problems. What actually happens is that the cam floats the valves and then the lifters automatically take up the clearance. But, lifter design has now reached such a state that we can grind almost

Top: Bob Joehnck, veteran hotrodder from Santa Barbara, California, built the engine which powered the Dees/Joehnck open-wheeled roadster to 223 MPH on gasoline fuel at the 1970 Bonneville Nationals. Aluminum heads are open-chambered ZL-1's, cam is a Racer Brown roller-tappet model with 0.650-inch net valve lift and 336° duration. Needle-bearing rockers are stabilized with a Jomar stud girdle. Note hydraulic unit on rocker cover for throttle actuation. Large-bore fuel injection is used. Ignition is stock Chevrolet transistor type. Crew standing behind car in lower picture includes Mark Dees, left, Bob Joehnck, center, and Dick Griffin. Joehnck does research and development for automotive companies and is usually deep in the heart of racing-engine construction for almost any kind of racing you can name.

any kind of a cam in hydraulic form—and do. In fact, we recommend them for most street and street/strip applications and even for boats and some circle-track racers. With a hydraulic cam you can forget all that valve-adjustment nonsense and leave your tool box locked up. And, the hydraulic cams are easier on the lifters. The GM-type lifters and our own private brand are self-purging because oil is always flowing through them to the upper valve train through the pushrod. Thus, there's no chance that air or dirt will be trapped in them to cause erratic action."

Racer continued, "Don't go wild on valve timing. Keep *effective* duration about 280 or 285 degrees as a reasonable maximum for wild street equipment with around 260 or so for sensible street applications, particularly on the inlet side. The exhaust can stand additional duration which will not detract from overall performance. Generally, valve lift should be in the range of 0.500 to 0.550 inch. Higher lifts show up well on this engine because the valves are not badly shrouded by combustion chamber walls, adjacent valves, etc. Longer duration can be used for competition, but even then the maximum should not exceed 300 to 310 degrees if RPM drops to 3,000 anywhere on the course.

"There is no *effective* timing until the valve is off its seat at least 0.015 inch to establish flow. It is easy to get added duration for advertising purposes by adding in the lengthy ramps which lift the cam gently off of its seat and return it the same way. When the *effective*-timing yardstick is applied, Chevy's street-mechanical and hydraulic cams show room for improvement. This is especially true of the hydraulic cams."

The "secret" of camshafts for additional performance is laid bare by lift vs. crankshaft-movement plots. Remember, too, that duration cannot be considered separately from overlap. We have included tables comparing the stock cams and another table comparing the offerings of one of the popular cam manufacturers. These are constructed on the basis of effective timing; that is, with the opening and closing measured with the valves approximately 0.015 inch off their seats.

Lift vs. crank-degree diagrams look similar for various cams because the designers must all work within the same limits for valve-train accelerations, spring pressure for returning the valve to the seat and compatibility of the tappets with the camshaft lobes under the spring loading, lubrication, and temperature conditions which exist. The rate of lift and descent, and therefore the curve shape, will turn out to be decidedly similar on various camshafts, provided the engineers know what they are doing. Be wary of vastly different curve shapes.

When comparing camshafts do not be impressed or swayed by a few degrees difference in timing or several thousandths lift. Such minor differences will not make two similar cams run that much differently.

If you plan to install a camshaft, order your cam and installation kit ahead of time, even if you are after a stock cam. Although Chevys are the most popular engines for reworking, there's always the chance that the particular cam you want was sold out yesterday.

Get the grinder's recommendations as to spring pressure so you will be able to install the appropriate springs.

CAMSHAFT COMPARISON TABLE

MODEL[1]	TIMING: IO-IC/EO-EC[2] LIFT/DURATION	APPLICATION NOTES
Hydraulic cams: All work without piston notching.		
SS-H-47	21-61/65-17 0.490/262	Works in any big block without piston modifications, smooth idle, good for about 6,000 RPM, works well with small-port heads and a 3310 Holley.
SS-H-48	31-71/75-27 0.500/282	Good to 6,500 RPM, lopey idle but will work with automatic transmission. Good ski-boat cam.
SS-H-35	33-73/77-29 0.560/286	Good to 6,800 RPM, works with automatic transmission if can stand lopey idle. Fine jet-boat cam.
ST-H-40	37-77/81-33 0.520/294	4-speed only. Street/strip cam.
Mechanical cams: These two work without piston notching in most instances.		
SS-19	30-70/74-26 0.490/280	Milder than Chevy's street-mechanical cam. Fine for heavy car with automatic transmission.
ST-8	40-80/84-36 0.520/300	Turbohydramatic with 4.1 or lower gear. Won't work with usual high ratios supplied with automatic transmission. Also good for 1/4-mile or shorter track use.
Competition cams: All are available with roller-tappet counterparts.		
STX-32	40-80/84-36 0.560/300	All-out drag with Turbohydramatic in full-bodied car 1/2-mile circle track, marginal for street, even with 4-speed.
STX-31	44-84/88-40 0.610/308	Drag super-stocker with 4-speed or automatic transmission. 5/8-mile or larger track. Good torque. Has been used successfully in some drag boats.
STX-27	46-86/90-42 0.610/312	Heavy car with 4-speed.
STX-43	46-86/90-42 0.640/312	Can be used on NASCAR or super-track if lack of bottom-end performance is not a problem. Has been used very successfully in drag boats.
STX-44	52-92/96-48 0.654/324	Maximum-effort cam, good for pro-stock used with tunnel-ram manifold.
STX-51	58-98/102-54 0.654/336	Supercharged boats or funny cars. Strictly maximum effort. Flat-tappet counterpart of roller cam used in Dees/Joehnck Bonneville roadster.

NOTES:

1. Cam models are Racer Brown. Table does not include all of his models and there are many other cam manufacturers supplying camshafts for the big-block Chevy.
2. Lifts are in decimal fractions of an inch and timing and duration are in crankshaft degrees.
3. Duration figured with hydraulic tappets at 0.008" lift; mechanicals at 0.015" at the lifter.

Chevy's own gear-drive assembly is used in marine and truck engines. It is impractical to convert to because a different distributor must also be used with a camshaft ground for reverse rotation. Plate behind cam gear bolts to block to hold cam in place. Gears are helical.

ROLLER-TAPPET CAMS

If you are building a big-block Chevy for street use, forget you ever heard about roller-tappet cams because you can spend that money for other items which are far more necessary. There's even some disagreement as to whether roller cams are required for an all-out racing big block, especially because the Can Am machines—which raced for big money—were not always equipped with them.

When lifts soar into the 0.650-inch region, such horrendous valve spring pressure—typically approaching 550 pounds open pressure—load a cast-iron cam and flat-tappet lifters to the point where a lot of wear can occur very fast. Multiply this spring loading by the 1.7 rocker-arm ratio and you have an effective net load of 925 pounds against the cam. This unit loading approaches 250,000 lbs/sq. in. where the lifter meets the cam lobe. For this reason, if no other, many drag racers and Bonneville machines are equipped with roller-tappet camshafts.

Owners of such machines are usually running the engines beyond the factory's rev-limit recommendation, often with nitro-methane-based fuel which reduces the oil's lubricating capabilities when the fuel dilutes the oil. For extreme engine speeds, the roller tappet allows using high spring loads, primarily because the roller tappet protects against destroying the cam lobe, assuming there is a reasonable shape. The tappets roll, helping the situation from a wear standpoint. The roller-tappet cam sometimes requires added spring pressure to hold the heavier tappets against the lobes. And the use of heavy spring loads and the extra mechanical complexity of the roller tappets with their guiding mechanisms increases opportunities for failures.

As a point of interest, we asked Racer Brown about differences in frictional loading in the camshaft and valve train—between the flat-tappet cam and a roller-tappet cam. He had not run any tests on the big block, but recollected that a small-block Chevy showed that the breakaway and sustaining torque required to rotate the camshaft was exactly half with the roller-tappet job. With the spring loads being run in the big block, there's a good chance that the roller would have an even greater advantage.

GEAR-DRIVEN CAMS

Ah, the romance of special high-priced racing components. Gear-driven camshafts are noisy, require a special high-cost ball-bearing distributor and are not necessary for street, drag-strip or most racing use. Save your money. You may see gear-drive cams being used in marathon race boats and in engines set up for long-duration racing, but there's no earthly reason to consider one for your street machine that gets rushed through the quarter mile a few times each month.

The stock nylon-toothed cam sprocket and matching chain works fine, assuming that you get a cam sprocket with concentric pitch line and hub.

Naturally, the people who make gear drives won't agree because they'd like to sell you a gear drive. Can Am machines manage to race with the Cloyes full roller chain. Those that don't use the Cloyes kit are using stock or replacement timing chains. This may be another of those places where you can save your money to buy the parts that you really need. Although the factory offers the 0.540/0.560 lift L-88 camshaft in a gear-driven version, that's the *only* one. The ZL-1 and late L-88 bumpsticks only come in chain-drive form, including the fuel-injection cam designed for the Can Am racers. If gear drives are so great, why didn't the factory continue in that direction? Think about that.

The stock Chevy gear-drive unit is a two-gear setup requiring a reverse rotation camshaft and distributor-drive gear. The racing-replacement gear drives are three-gear units, which means that the cam rotation is the same as a chain-driven cam. In fact, a chain-driven cam is used. Do not, repeat DO NOT, use a two-gear cam with a chain or three-gear drive or vice-versa because this can be very ruinous!

Gear drives have a machined part on cam nose to mate with minimum clearance into cover register. This prevents cam "walking" which can change spark advance as engine runs. Thrust bump button on chain-driven cam accomplishes same job. Non-Chevy gear drive may wipe out block thrust surfaces unless special oiling mods are made. However, these are three-gear sets so no special distributor is required.

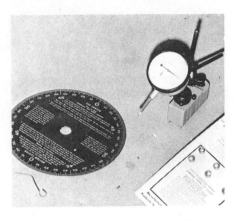

Three essential tools for camshaft installation in any high-performance engine are a degree wheel with pointer, a dial indicator with a magnetic base assembly, and some method for advancing or retarding the camshaft in relationship to the crankshaft indexing. In this photograph a series of offset bushings are shown. These are often used to advance or retard the cam.

Easy and accurate timing of the big-block Chevy requires a "degree wheel" to set timing. A machine shop can take care of this for you. Many speed shops sell a tape which wraps around the dampener to provide the same feature—at a fraction of the cost.

INSTALLING THE CAMSHAFT AND FINDING TDC

You will need a degree wheel or a degreed harmonic damper, a pointer, and a one-inch-travel dial indicator with clamps and brackets.

A slight modification must be made to two of your old valve lifters (tappet). The lifter needs a flat surface on which the stem of the dial indicator can ride for the following checks. A length of aluminum or steel rod can be fitted into the lifter body and a flat face turned on it. Don't use the radiused cup of the lifter as a contact point for the indicator stem, as it will cause inaccurate readings and possibly damage the indicator. Install the two modified tappets in the lifter bores for No. 1 cylinder.

Install the cam into its lubricated bearings in the block. Chevy suggests installing two capscrews in the cam nose to use as "handles" so that it becomes easier to install the cam without damaging the bearings with the lobes. Index cam sprocket onto the cam with the timing chain in place and install the cam-sprocket capscrews so valve-timing marks line up.

Install piston and connecting-rod assembly. Top and bottom rings should be used to hold the piston squarely in the bore, particularly if a dial indicator is used to find TDC. Attach the dial indicator so that its stem rests on top of the piston. Rotate the crankshaft slowly in its normal direction of rotation. When the dial indicator shows that the piston has reached its maximum travel, attach the degreed crankshaft pulley or attach a separate degree wheel to the stock pulley. Make an indicator tab from sheet metal or a piece of welding rod. Adjust the degree wheel and indicator to coincide with the TDC marking. Rotate the crank through an almost-complete revolution, stopping short of TDC when the dial indicator shows that the piston must still travel about 0.025-inch to reach TDC. Note the degree-wheel reading or make a mark on the crankshaft pulley. Continue rotating the crankshaft while the indicator passes through its maximum reading and backs down to 0.025-inch on the other side of TDC. Note reading. The ATDC reading should be the same as the BTDC one to show that you accurately located TDC. If not, start over again!

A large washer bolted to the deck surface of the block may be used to stop the piston to find TDC with the positive-stop method.

Another method positively locates TDC without using a dial indicator. It is so simple that it amazes us how few engine builders are aware of it. Bolt a strap across No. 1 cylinder. The strap should have a cap-screw in its center, placed to extend into the cylinder. Set the degree wheel and indicator finger at TDC by guesstimate. Rotate the crank by hand until the piston stops against the screw. Note degree-wheel reading. Reverse rotation and turn the crank until piston again stops against the strap on the opposite side of BDC. Note the degree-wheel reading. Move the indicator to a point exactly between the two readings.

This same method can be used when the engine is assembled and in the car if care is used. In this instance, the stop is made from a spark plug. Braze a piece of steel rod into an old plug base from which the insulator has been removed. The steel should extend about 1 inch from the threads. Remove all of the spark plugs from the engine. Install your special "plug" into No. 1 cylinder. Rotate the engine by turning the crank-pulley nut with a box wrench in one direction; with the fan belt in the other. Use extreme care so you do not damage a piston!

NOTE: The factory's timing mark on the crank pulley, when used in conjunction with indicator on the timing cover, is not a positive or accurate TDC location. Do not use it! There are no really reliable shortcuts for finding TDC or checking camshaft timing, so do not look for any. However, you can compare your accurate TDC location with the stock markings and scribe a correct line or move the indicator to provide a true indication.

DEGREEING THE CAM

A modern camshaft is a precise piece of equipment, but never assume that as such they are always correct in all respects. A camshaft that is absolutely perfect is impossible to find.

All cam grinders have established their own quality standards to ensure that you will usually receive an accurately ground camshaft. However, errors can and do occur in manufacturing camshafts and they cannot control the other parts in your engine.

Cam check is up to the engine builder and checking must be done in the engine in which the camshaft will live because *individual engines are different* and their differences can and do drastically affect valve timing. If the camshaft is installed in another engine, or is reinstalled after being removed for an engine teardown, the degreeing operation must be repeated. Maximum performance is not obtained by short cuts. The cam grinder can only control the camshaft, not the rest of the running gear in your engine. The engine builder must ascertain if there are any errors present in the relationship between crank and cam and it is he who must also correct any errors for maximum performance.

You can obtain maximum performance by making sure that the cam installed in your engine opens and closes the valves as intended by the cam grinder. And there is a second valid reason which makes cam checking essential: consider the hours of wasted labor if subsequent removal of the cam is required to correct valve timing! It's no trouble to make an error when indexing the marks on the crank and cam sprockets.

There is no doubt that you can install a stock or reground cam without checking. But, when you consider all of the tolerances working against you to cause less-than-perfect timing, why trust to luck? Some of the factors which affect timing include manufacturers' tolerances for the location of the keyways in the crank and cam sprocket dowels and their respective sprockets—plus the cam grinder's lobe-location tolerances. If, for instance, the location tolerance for the crankshaft keyway and dowel hole in the cam sprocket and cam nose is $\pm 0.75^\circ$, you could get a tolerance "stack-up" of 4.5 crankshaft degrees total timing error with every item off by the maximum amount. It is assumed that you have carefully followed the instructions and have installed the degree wheel and accurately located TDC. With the lifter on the center of the heel of No. 1 intake lobe, adjust indicator stem parallel to lifter travel in all respects. Preload indicator about 0.010. Rotate crankshaft and observe the indicator, watching for the point of maximum lift. Mark "IN" in pencil on the degree wheel. Rotate the crankshaft exactly one turn in the same direction until the pointer again aligns with your "IN" notation. This places the lifter in the mid-point of the cam lobe's clearance section. The cam has turned exactly 180° (half-turn). Without changing the indicator pre-load, set the dial-indicator face to zero. Rotate the crankshaft in the direction of running rotation and observe the dial indicator. When it shows that the lifter has raised an amount equivalent to the checking clearance, record the degree-wheel reading. This will be a certain number of degrees BTDC. If no checking clearance is specified or the cam tag has been lost, use 0.020-inch (close for most camshafts). Stock cams can be checked with 0.015 clearance for mechanical cams and 0.008 for hydraulic cams if you use the data in our comparison table.

Continue to rotate the crankshaft in the same direction until the lifter has risen up in its bore and fallen back again as it followed the opening and closing flanks of the cam lobe. Watch again for the indicator to reach the checking clearance or the arbitrary figure of 0.020-inch just mentioned. Record the degree-wheel reading as the number of degrees between the pointer and BDC. Add 180° to your two readings to get duration of opening measured at the checking clearance.

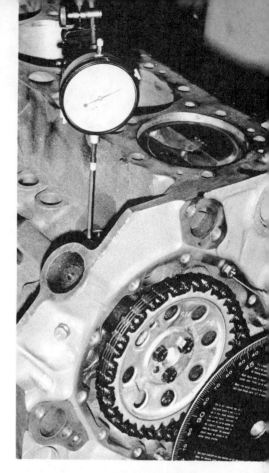

When taking a reading with a dial indicator off the top of a lifter, the plunger of the indicator must run in the same axis as the lifter and preferably on a flat surface created by modifying a lifter with an aluminum button.

Repeat the process several times to eliminate the possibility of errors. Each repeat check must be started at the "IN" mark on the degree wheel and the indicator's needle and dial face must be at zero. Cam lobes and lifter bores must be scrupulously clean and lightly oiled so the lifter can move in and out of the bore without binding and of its own weight. With everything right, the indicator will reach zero and stay there before and after the mark on the degree wheel is reached.

Transfer lifter and dial indicator to No. 1 exhaust and repeat the entire process, beginning with finding full lift, then turning 360 crank degrees and marking the wheel "EX"—using this as your starting point. Exhaust-opening checking clearance will be reached at a point BBDC. As with the intake, the number of degrees from the BDC mark on the degree wheel must be counted to get the opening point. Record this figure.

61

Degreed damper is a must for any serious engine work in a car on the dyno. Part of timing-chain cover has been cut out and used as an indicator for quick checks of the cam timing. This pro-stock racer's engine says that he is serious because these aluminum Can Am blocks cost many thousands of dollars each—if you can find one! Photo made at Edelbrock's prior to 1971 Winter Nationals. Note nylon bumper on cam nose.

Again, rotate the crank and observe the indicator as it indicates opening and then closing back down to the checking clearance. This will occur at a point ATDC, and the number of degrees from TDC must be recorded. Exhaust duration is calculated by adding 180° to the two readings recorded for exhaust timing. Again, repeat the process several times to eliminate the possibility of error.

Overlap is easily determined by adding the intake-opening degrees Before Top Dead Center to the exhaust-closing degrees After Top Dead Center. The entire cam-checking process presented herein has been taken with slight changes from Racer Brown's camshaft catalog which contains other data on correcting errors, etc.

Chrysler Corporation and some cam grinders use a different method of checking cam timing for high-performance engines. This is a quicker method but it is not consistently accurate unless checked at the lifter. It leaves the valve opening and closing points to fall where they may, but it is simpler because only one point is checked for each valve. Many builders check only the maximum intake-valve lift point when they are pressed for time. This is showing a lot of blind faith in the man who ground the cam, but suit yourself.

This method consists of finding the points of maximum valve lift in relation to the piston position of top center.

Assume the valve timing is: Intake opens 34° before top center, closes 66° after bottom center, duration 280°; exhaust opens 70° before bottom center, closes 30° after top center, duration 280°. To find maximum lift of the intake valve, subtract the intake-opening point from half the duration. Using the figures above, half of 280° is 140° and the intake-opening point of 34° must be subtracted from 140°. The answer—106 degrees—is the point *after* top center that the intake valve should reach maximum lift.

On the exhaust side, the point of maximum lift of the exhaust valve is found by subtracting the exhaust-closing point from half the duration. Using the exhaust valve timing of 70-30 as above, the 30° exhaust-valve-closing point is subtracted from 140°. The answer of 110° is the point at which the exhaust valve should reach maximum lift *before* top center. This assumes cam lobes are symmetrical on both opening and closing sides, and that's not usual!

To establish these points correctly, the piston position of top center must be found accurately, as in the previous method.

Finding the points of maximum valve lift is really just a repetition of finding top center, except that the dial indicator is transferred from the piston to the lifters.

This check can also be performed by measuring for maximum lift at a valve retainer in an assembled engine. The dial-indicator spindle must be parallel to the valve stem in both the longitudinal and lateral planes if the check is to be valid and one which will repeat.

This expensive dial indicator is graduated in inches on the linear scale and in thousandths on the dial—quite expensive, and certainly not necessary for occasional engine blueprinting. An indicator and base combination costing about $50 will do the job just as well.

VALVE TIMING CORRECTIONS

The previously outlined procedure—under ideal conditions—will show that timing points coincide precisely with those shown on the cam tag. But, due to factors already stated, variations can occur. A difference of one or two crankshaft degrees between actual installed timing and the cam-tag specifications should be considered the maximum-allowable error. A closer tolerance of ±1° (crankshaft) is even better.

If the timing check shows that actual timing is 'way off —10° or more—then the timing sprockets are probably misindexed by one or more teeth. Correct such a situation and rerun the timing checks. Racer says that if an engine is run in this condition, or if the valve-timing check is made at the valves instead of at the lifters, it is a foregone conclusion that you will end up with an engineful of bent valves at the least. And, you'll also get bent pushrods, damaged rocker arms, and possibly even a bent or broken camshaft. These are a few additional reasons why the timing check should be made at the lifters, as opposed to checking it at the valves, or not checking at all. Checking valve overlap with feeler gages at the valves by the split-overlap method is not recommended under any circumstances because of the errors which are usually present in the rocker-arm geometry. *Feeler gages cannot be used for accurate checks!* Correcting a timing error requires rotating the camshaft in relation to the crankshaft, which can be confusing. It is important to take care to advance or retard the camshaft and *not one of the sprockets!* Remember that the sprockets retain their relative positions; it is the shafts which must be moved. This is true in all instances except where the cam and crank sprocket are misindexed, in which case the sprocket/shaft assembly is rotated. Assuming that the timing sprockets are held stationary, the drawing illustrates the direction that the crankshaft or camshaft must be moved to advance or retard cam action.

Three valve-event drawings compare specification timing with advanced timing (all events early) and with retarded timing (all events late).

Advancing or retarding the camshaft from the settings given by the cam grinder is idiotic unless you have access to a dyno and will actually run tests in the various advance/normal/retard settings that you are advocating. While this sort of jockeying has been known to cause an engine to run stronger at one end of the RPM range—or the other—most experts run their cams on the stock marks. In general, advancing the cam increases torque and HP through low- and mid-range RPM. Retarding the cam kills excessive low-end torque and increases HP at the top end. When an offset key or bushing is used, repeat the entire degree check to ensure that the correction has been made properly, that is, corrected *the right amount*—in *the right direction*. A movement of 0.01396 inch at the crankshaft nose (inside bore of crank sprocket) adjusts crank timing one degree. Correcting a 5° timing error would require an offset key to offset the crankshaft sprocket 5° × 0.01396 = 0.070. Inexpensive offset keys or bushings can be used for making corrections, but, except for Racer Brown's, all are labelled in *camshaft degrees.* Keys or bushings can be made by any machine shop, but the readymade parts will be much cheaper. Rechecking timing after installing an offset key or bushing is essential!

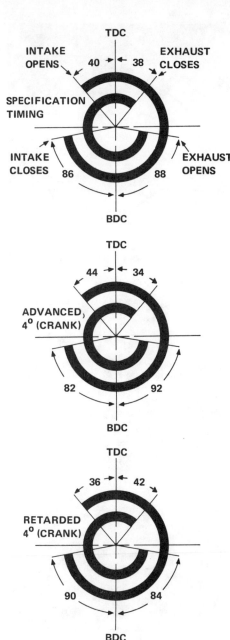

Comparison of normal specification timing with a cam which is advanced 4° (crank) and one which is retarded 4° (crank). 1° cam movement is equivalent to 2° crankshaft movement. Chevrolet's street-mechanical cam 3863143 used for this example.

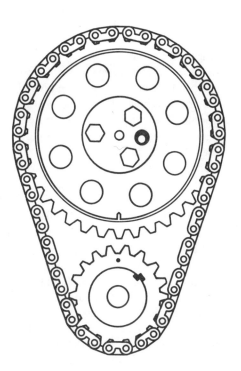

Drawing at left shows using bushing in the cam sprocket to advance cam. Offset key in crank sprocket advances cam when placed as shown. Reversing either retards the cam. Only one method needs to be used as various bushings and keys are sold.

In making valve-to-piston clearance check, clay is smeared in a thin layer over piston top. In lower photo, piston has pushed clay into valves, as described in text. A machinist's scale is being used to measure the clearance as indicated by the depressed clay on the piston.

VARIABLE TIMING DEVICES

Be wary of buying any device which alters cam timing *as the engine is running.* Make sure that you have personally witnessed a horsepower increase on a dynamometer before investing in such gadgetry. One of the authors observed a test wherein one of these adjustable wonders "adjusted" a 25 HP decrease from a strong-running big block. You didn't read about that in the ads, did you? Racer Brown tells of one of his customers who bought such a device. He gave up on trying to make it work as the ads stated and locked out the adjustable feature to allow making quick changes of the cam timing manually to compensate for different-length circle tracks.

Remember that advancing or retarding the cam can change the relationship of the valves to the pistons, perhaps reducing the clearance to less than that which is required.

VALVE-TO-PISTON CLEARANCE

When installing a high-performance camshaft from Chevrolet or *any* cam manufacturer for the first time, the clearance between the valves and pistons during the overlap period must be checked to determine whether adequate clearance exists between these components. This clearance should be checked if the cam has been running in the engine with success but other components are changed—such as rocker arms with a different ratio than original equipment, a stroker crank, a different piston type, or a larger-diameter valve head, or a longer connecting rod. A change in any of these components may cause the valve and piston to collide where the problem did not exist previously. Of utmost importance is the fact that just because a manufacturer says the cam won't push the valves into the pistons doesn't always make it true. He has no way of knowing what you have done to the engine to affect clearances . . . so don't expect him to buy new parts because of your haste or his assurance that piston notching would not be needed.

The checking of valve-to-piston clearance should be done only after the camshaft has been installed and degreed to the crankshaft as it will run. Clearances during the check will thus be the same as when the engine runs.

On a competition large-block Chevrolet with steel rods, the valve-to-piston clearance must be at least 0.100-inch for the intake valves and 0.125-inch for the exhaust valves. If the engine is to be run with aluminum connecting rods; increase the above clearances to 0.120 and 0.140 minimum, respectively, to compensate for expansion of the rods.

Something to watch for in an engine when the camshaft is advanced or retarded is the clearance between the valves and the pistons at their closest points. When a camshaft is advanced, intake-valve clearance at the piston decreases and exhaust-valve clearance increases; conversely, when a shaft is retarded, intake-valve clearance at the piston increases and exhaust-valve clearance decreases.

There are at least three methods for checking valve-to-piston clearance. The "feeler-gage approach" will be discussed first. With pistons, rods and crank installed in the block, lay on a compressed (used) head gasket and a cylinder head or just the cylinder head without a gasket. One cylinder of the head should contain an intake and an exhaust valve—held in place and in the closed position by Perfect-Circle or Raymond valve-stem seals. Install lifters, pushrods and rocker arms for this one cylinder. Adjust both valves for zero lash. Rotate the crankshaft one complete revolution to TDC. Lift each rocker arm up, push the valves against the piston head and measure the gap between the valve stem and the tip of the rocker arm. The measurement is the gap between valve and piston. If the measurement is made without a head gasket in place, add the thickness of a compressed head gasket to the measurement for an accurate indication of the valve-to-piston gap at TDC.

Check the valve-to-piston clearance for both valves in 5° steps to 30° or more on each side of TDC. Valves are sometimes closer to the piston when the piston is slightly away from TDC, so a single check at TDC is not a safe indication.

The "clay method" has been around for years, and offers a real advantage over the feeler-gage method because the valves may not be closest to the piston head at TDC. Lay two strips of modeling clay, each at least 1/4-inch thick, on the head of one of the pistons so that the strips are directly under the low side of the valve heads. Install the cylinder head which contains the valves for the cylinder to be checked. Place a used head gasket between the head and the block—making certain that the gasket is of the same type to be used in the final installation. A thin coating of oil on the valve heads will prevent clay sticking to the valves if they should touch. Secure the cylinder head to the block with capscrews tightened to the correct torque to duplicate final-installation conditions.

With the camshaft in the position that causes both valves for the cylinder to be closed, install the lifters, pushrods, and rocker arms for the cylinder and adjust the valve lash to .002 inch. Slowly and carefully rotate the crankshaft. Feel carefully for any tendency for components to bind or stop. If the shaft reaches a spot where binding is apparent, STOP! The valve may be hitting the piston, so don't use excessive force. Only a very light force should be required for the piston and valves to compress the clay, squeezing it out of the way. You should be able to feel what is occurring well enough not to damage anything.

Rotate the crankshaft at least two full revolutions, then remove the head carefully. Examine the clay. If either valve head touched the clay, slice the clay lengthwise with a sharp knife through the middle of the depression made by the valve. Peel one-half of the clay section off of the piston. The thinnest section of the clay can then be measured with a micrometer, vernier caliper or even a thin steel scale.

Although some "pros" might tend to look down their noses at the clay method of measuring piston-to-valve clearance, it is highly recommended for the beginning engine builder. First, this method gives the piston-to-valve clearance *accurately*. Second, a visual representation of the corrective cut is available so that you can use it as a guide to notch a piston. While experienced engine builders might not need this, it can be of immense help to someone who has never had to notch a hundred-dollar set of pistons.

To check valve-to-piston clearance by the "light-spring method" you'll need two light springs installed in a conventional manner on the intake and exhaust valves of the No. 1 cylinder. You don't need any trick springs here . . . fuel pump springs are lovely. Install a gasket and the cylinder head with two or three bolts. Install tappets, pushrods, and rockers. Install the camshaft to be run in the engine. Adjust valve lash to 0.002 inch. Rotate the crankshaft and stop at TDC with the intake and exhaust partially open in the overlap position.

Mount a dial indicator to the top of the cylinder head with the indicator "probe" resting at the top of the valve spring retainer. Set the indicator dial to zero. Now exert thumb pressure on the rocker arm until you feel the valve head touch the piston. By reading the dial indicator the valve-to-piston clearance may be determined.

To be double dead-sure about this, check the valve-to-piston clearance for both valves in 5° steps to at least 30° on each side of TDC. In some cases, due to piston approaching TDC and valve position, the valves will be in closer proximity to the piston when it is slightly away from TDC.

If the clearance is less than specified, the pistons will have to be notched or fly-cut to be run with that camshaft and rocker-arm set-up. Such notching should be done prior to balancing the engine assembly. Always notch the pistons to gain clearance. Never sink the valves!

ROCKER ARM GEOMETRY

In addition to installing the camshaft in the correct relationship to the crankshaft, you must observe another important point of engine construction: the relation of the rocker-arm-tip radius to the valve-stem center. If this is correct—with the centerline of the rocker's radiused tip coinciding with the valve-stem centerline at 50% lift—good

Measuring valve-to-piston clearance with a dial indicator. Only the experienced should attempt this without light springs. 425 HP 427 CID cast-iron head has the PC valve-stem seals installed.

things happen! Specification lift throughout the lift cycle of the cam is assured, side thrust of the valve stem against the guide bore is minimized to reduce wear and consequent oil consumption, friction is reduced, and higher RPM can be reached before valve float occurs. This is a point of construction which is often overlooked but it makes a vast difference in engine performance. Motorcycle race-engine constructors used and wrote about these things for decades; most auto-engine modifiers still refuse to realize that extra HP can be gained through careful attention to all details. Care produces more HP than a truckful of "black-magic" tricks. Bike builders often had only one cylinder so they had to make it produce maximum HP. Always watch current development in motorcycle engines and tuning procedures.

Getting back to the tip/stem relationship, there are several things which can affect this relationship: head milling, cam base circle, valve-seat location, valve-stem length, lifter dimension, rocker-arm differences and pushrod lengths. Some of these things become immediately obvious as sources of problems when we are working on the engine—then we shrug our shoulders and forget about the problems and go right ahead with the engine assembly, never realizing that in so doing we have seriously shortchanged the engine's breathing ability. Adjustments are possible through varying the height of the valve stem and/or the pushrod length; do not "sink" the valves!

Before continuing, let's look at some of the other aspects of ball-pivot rocker arms. Although these have virtues of light weight, low manufacturing cost, and simple installation, they also present problems and must therefore be considered a mixed blessing. The rocker opens and closes the valve with a continuously changing pivot-to-tip radius, moving through arcs of varying radii. With this changing situation, the rocker-arm ratio is highest at maximum lift, let's call it 1.7:1. It lessens as the pushrod moves away from the pivot, say 1.6:1. These figures, incidentally, are merely examples chosen to illustrate the problem and do not represent actual ratios which may occur. The problem is complicated by variances in the distance from the ball seat to the rocker tip. Just measure the valve travel with the cam at half and full-lift, using all 16 rockers, one at a time, on the same valve and cam lobe. Some produce more than 50% expected travel at half-lift, while others will show the opposite to provide more than specified valve travel at full-lift with less than 50% at the half-lift point. Obviously, you want 100% valve travel with the cam at full-lift position.

Make all of the valve-stem heights (*not lengths*) identical within ±0.005 inch (the rocker-arm-cover-gasket surface can be used as a reference point from which to measure heights). If any of the valves have to be shortened by grinding on the tips, reharden the ends by heating them cherry red and quenching in oil. Use a fairly large torch so the tip quickly gets cherry red without heat extending to the keeper groove. Hardening is wanted only on the tip. It is assumed that you have already seated the valves per directions in the section on cylinder heads. Reseating changes the stem heights, so do any seating *prior to establishing the heights.*

"Rocker-arm geometry is particularly important with the Chevrolet-type rocker gear. Nearly everything you touch affects rocker-arm geometry, usually for the worse . . . and nobody ever heard of it," says Racer Brown. "We do something with 427 Chevy pushrods and don't tell anyone about it. We make them *all* 1/16-inch shorter than stock to compensate for block decking, head milling, etc. It compensates in the wrong direction for facing valves and seats, but if it is necessary to make a deliberate 'error,' it's better to go in the direction of a shorter pushrod by far than to go for a longer pushrod 'by a little.' All this points out that rocker arm geometry, although extremely important, is usually completely neglected when building a Chevy engine.

"You can make individually tailored pushrods so that each valve, rocker and pushrod work as an integrated set to line up the center of the rocker-arm tip with the center of the valve stem at half-lift. But, this unnecessarily complicated procedure makes no provisions for broken or damaged parts. A better plan is like so: Get friendly with the local Chev parts dispenser or speed shoppe, depending on the brand and type of rocker arm to be used. Fill a bushel basket with the rockers of your choice and check them out in the engine using the *same* lifter, the *same* pushrod with the *same* valve on the *same* cam lobe.

"Set up the valve in the head with a light fuel-pump spring and set a dial indicator on the spring retainer. Install the rocker arm and pushrod and related bits and tighten the adjusting nut until the valve is off the seat, say by 0.002". Record the point of maximum valve lift with a chalk stripe on the degree wheel that's on the end of your crank. Rotate crank exactly one turn to the same mark to put the lifter in the center of the cam-lobe heel.

"Now very carefully measure valve duration in crankshaft degrees at 1/4, 1/2, 3/4 of full valve lift. Repeat the procedure on all rockers in the basket until, ideally, you get 16 rockers that are all exactly alike at all points. That would be ideal—and ideal things just don't often happen—so be prepared to accept at least eight rockers, plus a couple of spares, for the intakes. These should be as close as possible at all points. Ditto for the exhaust rockers.

"At this point, rocker ratio and rocker geometry do not mean anything. You are just trying to dig up an acceptable set of rockers, plus spares. Next, make up one intake and one exhaust adjustable pushrod and adjust the length of each to optimize rocker geometry. I prefer to do this with zero lash or close to it to ensure that the rocker arm is not cocked at an angle in relation to the valve-stem tip. At 50% lift, the center of the rocker tips should be in the center of the valve-stem tip.

"Rather than buying two adjustable pushrods for making this check, make an intake and an exhaust pushrod from stock parts. Braze a piece of a machine screw into one part and leave enough of the screw protruding so that it will accept a nut and extend into the other part of the pushrod for holding the two parts in alignment. The nut can be used to adjust the length.

"With the pushrod lengths established, it is no problem to get pushrods made to the required lengths (one length for intakes, one length for exhausts), again plus a couple of spares. We do it all the time. This plan gives interchangeability of pieces not obtainable in any other way.

"If the rocker geometry check is made as described, but with zero lash, the pushrods will be a very small amount too short, but again, this is better than having them too long because valve-stem side thrust and valve guide bore wear are minimized.

"Doing things this way more-or-less ignores actual measured rocker-arm ratio, but the end result is better functioning of the entire valve train and hardly justifies 'squeaking' about a few thousandths valve lift, gained or lost, from theoretical numbers."

Why not just use adjustable pushrods and save the bother of making special-length pushrods? Adjustables are made for the engine but they are somewhat heavier and could require higher valve-spring pressures. However, the idea of using adjustable pushrods in these engines is not all bad.

By now you are perhaps wondering why so many engines run as well as they do. You can begin to understand why the top tuners always run in front—but never brag about how quickly they can assemble a winning engine. Now that the need for careful assembly has been re-emphasized, let's proceed to the next item.

STOCK ROCKER ARMS

These stamped steel parts are one of the seven wonders of the mechanical world. They shouldn't work, but they do. They are lighter than any trick replacements that you can buy, cheap, and they seldom fail if you lubricate the grooved rocker balls when you install them. Stock rocker-arm covers with built-in drippers for lubrication help these parts live to a ripe old age. All of the rockers operate on grooved balls. Early engines were equipped with non-grooved balls, but all of the later ones are grooved. You can get these grooved balls at the Chevy store as P/N 3899622. These have a special heat treat, so get stock parts. The grooves hold oil in the high load areas to provide long life under the miserable high heat and marginal lubrication conditions to which the rockers are continuously subjected.

Never separate rockers and balls that have been run together. When cleaning, wire them together loosely so they won't get separated inadvertently. If you are assembling a new engine, try to get some used rocker balls from intake positions to use on the exhaust rocker arms. Why? Because the arms and used balls in good condition have already worn smooth and are less apt to fail during the critical run-in period. Exhaust balls always fail first if failure is going to occur. If your engine has an exhaust ball failure, move a used intake ball and rocker over onto the exhaust and install the new ball and rocker on the intake. Always coat the parts with E.O.S., Molykote or something similar.

Another Racer Brown tip: get the rocker arms and balls Parker-Lubrite treated and then paint each ball and ball socket with molybdenum-disulfide. Take these preventive measures and Racer Brown claims you'll have zero problems with rocker-arm wear.

Rocker-to-stud clearance

It is essential to check the clearance between each rocker and its stud with the rocker at full lift. If a straightened piece of paper-clip wire can be slipped between the stud and the front of the rocker slot edge, everything is fine. If not, elongate the slot by grinding. The rockers with an "H" on them have a longer slot than the "L" rockers. The "H" rockers are P/N 3959182.

Roller-tip needle-bearing aluminum rocker arm compared with a stock Chevy part. Stock piece represents the simple, direct and extremely reliable approach used throughout all Chevrolet engines. Correctly installed stock rockers are adequate for almost all types of racing. Their use can help to reduce your engine-building costs drastically!

Cam thrust buttons are sometimes used on the cam nose. This needle-bearing type sells for about $5 and works against a 1/8-inch-thick steel pad brazed inside the cam cover. About 0.008-inch end clearance will end spark scatter problems because the end button will keep the cam from walking back-and-forth in the block. Roller cams without "straddle"-type tappets are the worst offenders in this regard and an end button should be considered absolutely essential for any engine so equipped.

Don't reface stock rockers — The thin case hardening of the stamped-steel parts will not tolerate regrinding the tip radius without seriously impairing reliability. Rockers with worn tips must be replaced!

Roller tips on these rockers are so small that no rocking "couple" can be developed between valve stems and tips to cause the advertised rolling action to occur. Can Am engine builders consider these to be throwaway items after about 2,000 racing miles.

Adjusting screws on needle/roller rockers often prevent installing stock covers. The 1/2-inch aluminum spacer here was cut with a saber saw, using a gasket as a guide.

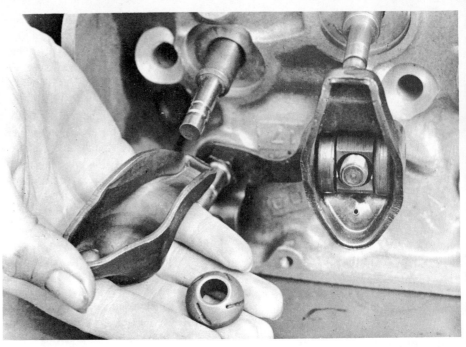

Stock big-block rocker with its grooved rocker ball is compared with Boss 302 Ford needle-bearing rocker installed on an aluminum ZL-1 head. Special nut is used to hold the Ford rocker onto the rocker stud.

NEEDLE-BEARING ROCKER ARMS

Do you really need needle-bearing rockers on a street-driven car? Probably not! A lot of money is wasted every year by enthusiasts who whip out their checkbooks before putting their minds in gear. Somehow, the possibility that super-expensive trick parts *may not be required* never seems to sink in. Parts manufacturers and speed-shop owners love their customers for their inability or refusal to think about what is really necessary for their engines.

Your street-driven engine is not launched from every stop light with the RPM approaching 8,000 as are the engines of pro-stock drag racers. You are in a different ball game. And, we might mention in passing that a lot of the pro-stockers still use the stock rocker arms, especially those that don't get the parts "laid on them" for free. Your street machine is not equipped with a dry-sump system that needs reduced circulation to the rockers so that the sump can be kept adequately pumped out in long-distance events.

Here are some facts which you may have overlooked. Needle-bearing rockers are often used because they reduce the side loading on the valve stems, thereby reducing friction and perhaps reducing heat. This is the single positive reason for using these rockers in anything other than a Can Am car. Needle rockers aggravate stud breakage to say nothing of the fact that the rockers themselves break. That's another fact that does not get into the ads or data sheets. This makes it essential to use special rocker studs or to shotpeen the radius at the top of the stud base. And, the top of the stud must be machined flat to allow using the setscrew inside of the adjustment nut used for these rockers. The stud-breakage problem could lead you to buy a stud-stabilizing device such as the Jomar Stud Girdle. Now you have doubled what you had planned to put into your engine when you started out to add the special rocker arms.

As with stock rockers, not all of the needle-bearing types provide full lift. The ratio can be less than 1.7:1 and you'll have to install a set in an engine to find out whether the type that you are thinking of buying gives full lift. Further, the longer rocker-adjusting nuts may interfere with the rocker covers so that you have to use double gaskets or shims for installation.

When using the needle-bearing rockers, the edge-orifice mechanical lifters should be used to reduce the flow of oil to the heads.

If you have had problems with your stock rockers, check out that rocker-arm geometry before throwing brick bats at the stock parts. Getting the geometry correct—as described elsewhere in this chapter—could solve your problems for a lot less money than you had been planning to spend. Stock parts nearly always get the job done if you'll think about what's happening and remedy the problems which have been caused by deviating from stock combinations.

Most needle-bearing rockers are aluminum and therefore have no fatigue resistance. They have been known to fail. One builder claims six hours as a maximum "safe" life for an aluminum rocker. George Bolthoff and Bill King both suggested that a maximum of 2,000 racing miles could be run before replacing them. And, the roller tip on these arms is not particularly helpful regardless of what the ads say. So, you might be better off with steel rockers.

TIMING CHAINS & SPROCKETS

Big-block timing chains and sprockets are not all alike. Three styles have been supplied directly from Chevrolet. First was a 3/4-inch-wide Link-Belt chain used in all 1965-66 models. Chain 3860036, with crank and cam sprockets 3860035 and 3856356, was thrown onto the heavy-duty parts pile when 5/16-inch pushrods were installed in hydraulic-lifter engines as part of a noise-reduction program. The Morse chain subsequently used is 5/8-inch wide. The third chain is sometimes referred to as a roller chain, even though the split bushings do not roll. It is used in some truck applications.

It is generally accepted that the wider Link-Belt-style chain is the one to use for high-performance applications because it is less prone to stretching than the skinny chain. See caption, right.

If you are building an all-out racing engine, consider the Cloyes Tru-Roller chain 921-1013. It is stronger than the stock chain and less prone to stretch than any of the stock chains. The chain is made for Cloyes by Reynolds, the famous English firm, and the solid-bushing rollers really *roll*. Cloyes guarantees that this chain eliminates chain-stretch problems and states what the timing error due to stretch or slop will be when the chain is installed. In the case of the big block, it is 3/4° at the cam, or 1.5° crankshaft, as compared to a typical 3° stretch error common with split-bushing-type "roller" chain sold as a truck replacement by Chevrolet and touted as roller chain by other timing-chain suppliers. Cloye's cam sprocket is Tuftrided and Magnafluxed cast iron, dynamically balanced to 1/8 inch ounce at 5000 RPM. That's equivalent to 10,000 engine RPM! The pitch diameter of the sprockets is larger to accommodate the large rollers used in the Reynolds chain, so you can't buy just the chain alone to work with your stock sprockets.

In the accompanying photo, you'll note that there are three keyways in the crank sprocket. This allows the cam to be installed in the standard setting, 4° advanced or 4° retarded.

Three Chevy timing chains. Left is split-bushing type used in some trucks. Center is Link-Belt type used in hi-perf engines since '67 and in all '65-'66 models. Right is narrow Morse chain typical of low-perf big-blocks since '67 model introduction. Cast nylon teeth are often non-concentric to sprocket hub creating wierd cylinder-to-cylinder timing variations.

Three keyways in Cloyes Tru-Roller crank sprocket allow installing cam in stock position or advanced or retarded by 4°. Drill indentations in cam sprocket are made during dynamic balancing of this part — another plus that makes these sets worth their price. Below — Tru-Roller's wide teeth and bushings compared with Chevy Truck set at right.

If you have had your block line bored and the cam-to-crank center distance has been reduced approximately 0.005 inch, Cloyes offers sets on special order with sprockets having larger pitch diameters to compensate for the closer positioning of the crank and cam.

Cloyes also offers a high-performance split-bushing timing-chain set which uses the same type of steel crank gear with three keyways, but without the no-stretch feature of the Tru-Roller set.

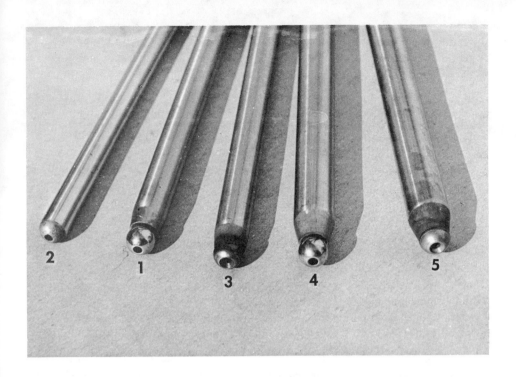

Shown here approximately lifesize are five pushrod styles which have been used in the big blocks. Text describes these by numbers shown in photo.

PUSHRODS AND PUSHROD GUIDES

Three diameters of pushrods are used in big blocks. Pushrod diameter must always match the guide plate. The guide plate holds the pushrod to guide the rocker arm in an arc which matches valve angles. If the large pushrod guide plate were to be used with small pushrods, guiding would be lost and you would end up with wrecked pushrods, rocker arms and valve guides, plus you'd probably find the ends of the valve stems mushroomed beyond repair. Do not attempt to modify small pushrod guides to fit larger pushrods.

Five different kinds of pushrods have been used in the big-block Chevy engines. Because intake and exhaust are different lengths, at least 10 different pushrods have seen service in big-block applications, not counting truck engines with the raised decks.

1. 1965-66 - All engines had pushrods with steel balls welded onto tapered ends of 3/8-inch diameter tubes. 1967-70 - Same pushrods used in solid-lifter engines. Intake 3864907; exhaust 3864908. Discontinued when the pressed-in design was introduced.
2. 1967-71 - Hydraulic-lifter engines used a 5/16-inch-diameter tubular steel pushrod with closed ends, similar to pushrods used in small-block engines. Intake 3904376; exhaust 3904377.
3. Replacement for the first pushrod for use in solid-lifter engines of lower-HP varieties. 3/8-inch diameter with ends pressed into the tubular pushrods. Intake 3946067; exhaust 3946069. Introduced in 1970.
4. First design L-88 engines had 7/16-inch-diameter tubular-steel pushrods with the steel-ball ends. Discontinued when pressed-in design introduced.
5. Second design L-88 engines, ZL-1 engines and LS-7 engines have 7/16-inch-diameter tubular-steel pushrods with pressed-in ends. Intake 3942416; exhaust 3942415.

Any of the big-block pushrods can be used. Make certain they are straight before you start dropping them in. If several are bent, throw the entire set away and replace them with the newer, heavy-duty 7/16-inch pushrods and guide plates.

When you ask for the large pushrods by part number, don't assume that you are being given the correct ones when they are laid on the counter. Measure them and eyeball them to make sure that you are getting the newer type with the pressed-in ends. More than one parts man has been guilty of cleaning out his old stock first. You don't want the ones with the punctured ball bearings fitted to each end because the balls can pound into the tubes or come loose. Get the late ones.

Although you might think that installing pushrods is a simple drop-in procedure, be sure to read the details in the Rocker Arm Geometry Section of this chapter so that you will understand how the correct pushrod length can add HP to an engine.

Top of rocker stud has been ground flat to provide seat for locking-type nut and setscrew used on needle/roller rockers. Pushrod guides should be deburred and radiused wherever an edge contacts the pushrod. Guides are hard, so this will require mounted stones in a grinder.

Two lifter types used with mechanical lifter cams. Inertia-flapper type (left) is called "piddle-valve." Edge-orifice (clearance-orificing) lifter should be used as is in engines with needle-bearing rockers, but 0.003 to 0.005-inch flat should be added between hole and groove (arrow) if engine has stock rockers. Edge-orifice lifters reduce total oil circulation by 10 to 20%—especially helpful for dry-sump oiling or restricted oil-pan capacity. These lifters usually eliminate any need for restricting the lifter galleries. Racer Brown offers pushrods with restricted oil flow if further help is needed to cut down oil circulation.

TAPPETS (valve lifters/cam followers)

Regardless of what you call these parts that convert the rotary motion of the cam lobes to a linear motion for the valve opening and closing, they are extremely important . . . and deserve more than a passing mention.

The camshaft and lifters must be "compatible" or extremely rapid wear results. Why? Because the force of the valve spring, multiplied by the rocker arm, works to keep the tappet in contact with the cam lobe. This force squeezes out and wipes away or vaporizes the lubricant, thereby promoting wear. Unless the metal types of these rubbing parts are exactly compatible in metal composition, hardness and surface finish, very fast wear results.

Tappet faces are not flat, incidentally. They are slightly convex with about a 30-inch radius. This is done to make the tappet turn as it works against the camshaft, thereby constantly exposing a new surface under the load so that the cam lobe and tappet will not wear out in a hurry. So, don't face tappets flat with a valve facer or you'll be rebuilding your engine. You can quickly check whether tappets have the desired curvature by holding the faces of two tappets together. They should "rock." If they have "gone flat"—get new ones.

If tappets have a few small pits in them—without any swirl marks, radial lines or other evidence of excessive wear—it's o.k. to use them again. But, you'll have to install them "back home" against the same lobes of the old cam, right where they ran originally. If you are installing a new cam or a reground cam, spend the money for a new set of genuine Chevrolet lifters (or use the lifters recommended by the cam grinder). You'll be glad that you did. Lubricate the cam and lifters carefully as described in the engine assembly chapter. Hydraulic tappets are highly recommended, even for high-performance engines.

There is some disagreement as to whether the oil supply to the lifters should be restricted from the rear main bearing to the rear cam bearing—which also restricts the flow of oil to the lifters. Chevy engineers we talked to insisted that the edge-orifice mechanical lifters are an adequate restriction and suggested that no additional restrictions need be installed. Some big-block engine builders suggested tapping the passage for 1/4-20 and installing a socket set screw drilled to 0.090" as a restrictor jet. George Bolthoff of Racing Engine Systems Development in Santa Ana, California formerly headed McLaren's engine-preparation group in England. He told us, "We tried restricting the oil to the lifter galleries but finally quit doing it because it was just one more of those trick things that take time but are not needed on the big block." Thus, we'd suggest that you run the engine in its stock form with the edge-orifice lifters prior to making any modifications. Then, if it appears that there is too much oil "upstairs" in the rocker covers, you might try restricting the supply. This would only be allowable with needle-bearing rockers.

Two Chevy mechanical valve lifters are available for the big block. One, P/N 5232695, appears similar to a hydraulic lifter and controls oil going out the top of the lifter with an internal flapper valve actuated by inertia. This is the most common type of big-block Chevy lifter and it's also used in the Z-28 version of the small-block engine. Lifter No. 5231585 is found in the 327 engines manufactured from 1959 to 1965 and also some of the later 302 engines. The lifter controls oil moving out of the top on the basis of the clearance available between lifter and the bore. Using this lifter reduces the total oil circulation rate into the heads by as much as 20%—which should be considered if you can't easily add to the capacity of the oil pan in your vehicle or plan to install a dry sump oil system on the engine.

Although it is a commonly used trick among the experienced big-block builders, Jerry Thompson of Troy Promotions was the first to tell us about what to do to the lifter bases to get better longevity and to ensure that the cam has the best possible chance of living—especially during those first few critical minutes right after the engine is fired up. The factory finish on new tappets or lifters is "too aggressive" so Jerry and other big-block builders recommend that you prepare the tappet faces by chucking each tappet in a lathe and running the lathe at high speed. Hold a piece of solvent-wetted No. 600 Wet-or-Dry abrasive paper against the tappet base very lightly so as to "dust it off." Do not try to hand rub the tappets against abrasive paper supported on a flat surface or you could destroy the spherical contour of the tappet bases. Thompson also suggests that the best reliability combination is a used cam with used lifters which have been given the abrasive-paper treatment just mentioned. Next in order of reliability or freedom from trouble would be new lifters, treated as recommended and installed on a new cam. "Scariest of all—and most failure prone," according to Thompson, "are new lifters installed onto a used cam, even if the lifters have been correctly treated and the correct lubricating techniques are used."

HYDRAULIC TAPPETS

Stock Chevy hydraulic tappets, P/N 5232670, are recommended for your stock or reworked engine which is equipped with a hydraulic-lifter camshaft. These lifters, when preloaded by turning the adjustment 1/8 to 1/4 turn in from clatter, will operate faultlessly beyond 7,000 RPM if the cam lobe shape is right. Chevy expert Larry Eave points out that hesitation and rough idling which often accompany turning in the adjustment from clatter can be avoided by backing off the nut until the clatter occurs, then just tightening the adjustment nut enough to eliminate the clatter. Do not establish the preload at that time, but continue to the next valve. Adjust it the same way and continue through the entire valve mechanism. When all of the valves have been set at the "just-not-clattering" point, shut off the engine. Now establish the 1/8 to 1/4 turn preload at each adjustment nut. It works!

Never mix your tappets if you are reinstalling the old tappets on your old cam. Keep each tappet with the cam lobe on which it has been running or replace all of them. You can precondition the lifter bottoms with abrasive paper as described in the previous section.

Avoid the so-called fast-leak racing tappets as these are known to collapse at both idle and high RPM.

2.3-inch valves are *not* the answer! Avoid going after the last little bit when it comes to intake-valve size. Although Chevy makes and sells a 2.3-inch valve, even fuel-injected 495-CID Can Am engines don't appreciate this extra opening! So hold off. This is one place where *smaller is better.*

ROCKER ARM COVERS

Rocker arm covers might not seem like a part of the valve mechanism, but they can interfere with valve action. When using a high-lift camshaft with needle-bearing rocker arms, bolting the valve covers on may open a valve or the adjusting nuts could hit the built-in oil drippers. Don't think for a minute that this kind of problem hasn't created its share of frustrated mechanics. Spread machinists' blue dye on the inside of the stock covers before installing them so you can check whether the rockers are hitting. Look for bright spots when you pull the covers off after turning the engine through several revolutions.

If there are places where the rockers contact the covers, place the cover over a socket and dimple it slightly with a ball-peen hammer. With solid aluminum covers such as those made by Edelbrock or Mickey Thompson, you'll have to resort to a rotary file to make clearance. If the oil drippers have to be modified for clearance, be sure to leave the exhaust drippers intact if you are using the stock rocker arms. If you cannot do this, get other covers or use your stock ones. If the interference is severe, double gaskets or spacers may be required to get the covers away from the adjusting nuts on needle-bearing rockers.

OVERSIZE VALVES

Installing bigger valves has been mentioned in the cylinder head section. Is this worth the time and money? It most certainly is, especially when you are installing a special camshaft. Larger valves help you get your "money's worth" in performance. The combination of big valves and a special camshaft will add HP to any engine if properly installed and tuned.

The gain in power afforded by larger valves is attributable to the added mixture-flow area (circumference x lift) when the valve is opened. With a larger-than-stock valve, the area is greater at every portion of the lift curve. Bigger exhaust valves are equally important in making these engines breathe, especially at high RPM. Regardless of the type of cam you choose, install the 1.88-inch exhaust valves and the 2.19-inch intakes in your heads if your budget permits.

SPECIAL LIGHTWEIGHT VALVES

TRW makes titanium intake valves which weigh a mere 82 grams each instead of the 139 grams which is typical for the steel 2.19-inch intake. And, they also make hollow-stem exhaust valves which weigh 136 grams each. Manley sells the same components.

Pro-stockers might consider these valves as insurance against valve float at high RPM. Or, the valves could be run with less than the usual valve-spring force so long as valve float was avoided.

Due to their cost and consequent limited market, titanium valves are made in small quantities with exotic production processes. Because so few are made, they may not be subjected to the same critical controls applied to bread-and-butter items made in the same factories. Thus, slip ups can occur. These are racing parts and there is no guarantee on them. *Racing costs money...how fast do you want to spend?*

RETAINERS

Valve-spring retainers are another of those romance items where the good stuff in the HD Parts List is often overlooked because the enthusiast engine builder is blinded by all of the trick multi-colored aluminum retainers offered by the cam grinders. General Kinetics' catalog tells it like it is, "High-strength steel retainers are recommended whenever the engine is to be driven at high RPM over an extended period of time such as in sports-racing or dual-purpose cars. If aluminum retainers are used they should be inspected and replaced periodically."

A major point of erosion is usually found where the flat damper coil contacts the aluminum retainer. If you see aluminum specks floating around or deposited on the head when you remove the rocker covers, this indicates that the retainers are being eroded by the springs. The oil may look "glittery."

While not all cam grinders will agree with us, we'll stick our necks out and say that you won't be wrong in using valve-spring retainers, preferably the best stock ones, P/N 3989353. This recommendation applies to any and all uses except perhaps pro-stock drag racing engines where the need for higher revs could make it worthwhile to spend the money required for a set of Mr. Manley's *titanium* retainers.

As a general rule any valve-spring combination that *does not* have a flat-wound damper can utilize aluminum retainers successfully. That damper is *the culprit* when it comes to a chewed-up aluminum retainer.

VALVE SPRINGS

It is not unusual to hear of an engine builder installing heavier valve springs without making any other changes to the valve train. There is no sense in running more valve-spring pressure than is absolutely required because this just takes more HP to turn the cam over in its bearings. Don't buy the all-out triple valve springs for your high-lift hydraulic or your street-mechanical camshaft. Use the factory recommendation for the camshaft—nothing more—nothing less!

The ultimate spring which has been released by Chevrolet includes a damper

Three of the valve-spring/retainer/stem-seal combinations offered by Chevy for the big-block engine. 1 - 3859911 single outer spring with flat damper and rubber umbrella seal. This is not recommended for a hi-perf engine because the spring has a bad habit of breaking when subjected to high RPM, mainly because it wasn't designed for such use in the first place. 2 - 3970627 dual spring includes a retainer with nylon seal attached. This has purple markings and the yellow end (closed coils) goes toward the head. 3 - 3916164 dual spring with damper uses retainer 3879613 with small stem umbrella built in. Spring is marked with silver and brown paint, has 1-7/16" OD. Diameter of 1 and 2 is 1-17/32". The No. 3 spring combination shown here was the 1st design L-88/ZL-1 spring and it has been replaced by 3989354, which is separately shown. The No. 2 spring will be found in most late-model high-performance engines. In general, you should follow Chevrolet's own recommendations for valve-spring application. Do not run more spring pressure than is absolutely required.

and an inner spring, 3989354. It requires the previously mentioned lightweight steel spring retainers, 3989353. Install the springs with 125 pounds seat load. Typical installation height is 1.85 inch. The springs have been used successfully with aluminun retainers.

Measuring Valve Springs

An accurate spring-testing fixture is necessary for these tests. Your local garage or parts outlet may have one. Look until you find one. You'll also need installed-height specifications to test the springs.

Note that the inner spring sits on a step in the retainer. Using the retainer when testing the springs simplifies the job. Remove the springs and place them in the spring tester. Then, compress them to the installed height and note the pounds of force exerted on the seat.

Now simulate valve opening by compressing the spring assembly to full lift and read the open pressure. The importantant thing here is to watch for coil bind, or solid stack, when the valve is fully opened. The springs should have at least 0.060-inch travel beyond the full valve lift for a margin of safety.

Tension on all valve springs should measure within 10 pounds for use in a high-performance engine. Every spring must be tested, even brand-new ones.

When you have established that the springs are within specification, cut a length of welding rod to the installed height for the outer spring. This will be 1.88 or 1.90 for the stock high-performance springs. It may be different for springs provided by a cam manufacturer. Install the valves in the guides and add the retainers and keepers without any springs.

Pull against the retainer to hold the valve against the seat. Insert the measuring rod between the valve-spring seat and the retainer. There will probably be a gap: the rod will be shorter than the

Bullet-proof spring for racers: 3989354 has an outer, inner and a damper spring. Note that the damper does not go all the way to the top of the spring in the non-loaded condition. Steel retainer does not include umbrella seal, so stem seals are supplied as 3998265. These seals install without machining the guides. Who has the better ideas?

VALVE SPRING TABLE
(pounds at height)

	L-88 ZL-1 3916164 1st design (triple)	L-88 ZL-1 3989354 (1) 2nd design (triple)	454 3970627 (2) dual
Outer			
Closed	75 @ 1.88"	88 ± 6 @ 1.90"	75 @ 1.88
Open	192 @ 1.32"	226 ± 12 @ 1.30"	243 @ 1.38
Inner			
Closed	41 @ 1.78"	40 ± 4 @ 1.80"	30 @ 1.78
Open	101 @ 1.22"	100 ± 9 @ 1.20"	86 @ 1.28

(1) Specs are for inner and outer only and do not include damper contribution. Combined loads closed are 128 lbs.; 326 lbs. open at 0.600-inch lift. Add approximately 25 lbs. for damper contribution. Cap 3989353 to fit spring does not include a seal, so Chevrolet 3998265 valve-stem seals must be installed. Outer spring is identified with a blue stripe and a grey stripe.

(2) This spring combination is an across-the-line replacement for 396, 402, 427 and 454. It includes a retainer with umbrella-type oil shield for the valve stem.

space between the spring seat in the head and the retainer ledge. Insert valve-spring shims between the end of the rod and the retainer to take up the gap. Remove the retainer and *install those shims over that guide.* Mark that retainer so that it can be reinstalled on that valve. Do this for each of the valves. Unless the springs are installed at the correct height, you will not have the correct seat/open pressures and valves may float 'way before the desired RPM has been reached.

In general, you can figure that springs which have run for more than six months in a passenger car will not be usable for an engine which is to be turned to specification RPM.

Spring-retainer-to-valve-guide clearance
Make this check before installing the springs, but after any seals have been installed. A machinist's scale can be held alongside of the retainer as you move the valve to its specification lift. There must be at least 1/16 to 1/8-inch clearance between the valve-guide end or installed valve-stem seal and the spring retainer when the valve is in the open position. Valve guides may stick out of the cylinder head far enough to interfere with the spring retainers if a high-lift camshaft is also being used. It may be necessary to machine valve-guide ends when a high-lift cam is installed. Most valve guides are installed with adequate clearance for the stock low-lift hydraulic cam.

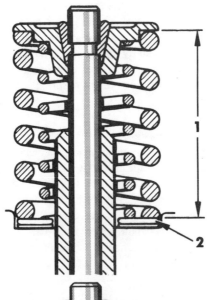
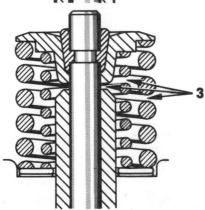

1. Installed height must be correct for each spring.
2. Valve springs seat on heavier shims, thinner shims go against head.
3. With valve at full lift, inner/outer spring coils should have 0.012-inch clearance between each coil; retainer and guide—at least 0.060 inch.

7/16-inch diameter pushrods from Racer Brown. Light-colored metal ends of the pushrods are hardened to work against the pushrod guides. Pushrods are available unassembled for those who plan to tailor each pushrod to provide exact half lift of the valve at half lift of the camshaft. Triple spring has heavy outer, damper and inner spring to keep control of the heavy valve train. Two styles of hard-anodized aluminum retainers fit snugly into springs to avoid wear in competition applications. Retainer with the slightly dished-in top allows setting a different spring height than the flat-topped one.

By constructing the tray from light-gauge, soft aluminum the tray may be trimmed out with tin snips or a saber saw. Top Chevy-engine builder, Lance Morris, advises saving time on the project by making a template of light cardboard and transferring the pattern to the metal.

"Grumpy" Jenkins uses a metal plate in the very bottom of the lifter gallery to contain the lifters if some of the valve gear should let go.

A home-made splash tray in place shows small tabs which locate the tray between the heads. Soft aluminum about 0.050 thick was used for this tray.

. . . then another metal plate is dropped into place in an effort to keep hot oil from the bottom of the intake manifold.

Two of the Jenkins' lower baffles showing one in place. Other baffle is shown inverted atop block so construction will be obvious. Slots are provided for the pushrods. Five screws fit holes drilled and tapped just above cam-bearing centers in this ZL-1 aluminum block. Photo courtesy Popular Hotrodding.

65-66 engines use a grooved rear cam bearing 3876863 (not shown) in conjunction with a grooved rear cam journal as shown on the end of this cam. 67-and-later engines use a non-grooved rear bearing, 3967810, as shown here. These engines also use a cam with a rear journal that is not grooved.

GROOVED CAM JOURNAL CAUTION:

1965-66 big-blocks require a grooved rear cam journal and a grooved rear cam bearing with three oil holes: two 0.250-in. holes and one 0.313-in. hole. The other four cam bearings have a single 0.116-in. oil hole. The combination was used to provide valve-lifter oiling on these early models. In 1967 and on all subsequent models, the lifters are oiled through an annulus groove behind the rear cam bearing. The rear bearing is *not* grooved and *all* bearings have a single 0.116-in. oil hole. If you have to use a grooved cam with a 1967 or later block, don't worry. Internal oil leakage will *not* be excessive if the '67 and later rear cam bearing is used.

If you are building a 1965-66 block, use the No. 5 (rear) cam bearing with additional holes and use a *grooved* cam to ensure an adequate oil supply to the lifter galleries.

CAM BREAK-IN TECHNIQUE

Most camshaft wear occurs in the first 30 minutes of operation. Professional racers are known to set up their engines with lightweight single valve springs, douse the lifters and cam liberally with E.O.S. and then overfill the crankcase. The engine is then run for 30 minutes to an hour at not less than 2500 RPM with zero idle-speed time.

New valve-train parts go through quite a change in lash during run-in and the lash should be checked frequently until it is stabilized. Remember to check the lift at the valves to determine whether any of the cam lobes are experiencing undue wear.

Valve lash is rechecked after the fast-idle run-in and any excessive lash is carefully noted so that cam lobe can be measured for lift before running the engine again. If the cam lift is not up to specification, the lobe has worn and it is time for another camshaft. This can and does occur. If the cam has not worn, then the valve springs are changed for the correct ones and the crankcase is drained and refilled with racing oil.

Use a fitting which screws into a spark-plug hole in conjunction with compressed air to hold the valves against the seats while changing the springs without removing the heads. Valve spring compressors which work in conjunction with the rocker-arm studs are available.

The run-in procedure just described is especially recommended for flat-tappet camshafts which are to be run with very high spring pressures.

HOW TO SET VALVE LASH

1. Stabilize the oil and water temperature by idling the engine for at least 30 minutes for a cold engine and about 10 minutes for an already warm engine. While the engine is stabilizing, remove the rocker-cover bolts and whatever else is necessary so that the rocker covers can be removed rapidly.

2. Shut off engine. Remove the most difficult to remove rocker cover. Inspect for burned balls or broken valve springs. Run a magnet around the head and inside of the rocker cover to pick up any chips, etc.

Start engine and lash the valves to the specifications shown in the table. Lash should be adjusted for a snug fit on the feeler gage. The rocker arm must not be cocked to one side or a faulty lash will be the result. The feeler gage should be inserted and removed straight in toward the rocker-arm stud—NOT pulled from the side of the valve stem. The feeler gage should be free of burrs on the end, of course. If a feeler gage is used to check clearance while the engine is running, consider that gage as a throwaway item after lashing the valves one or two times. When correctly lashed, the valve train should have only a slight mechanical noise at idle and the engine should idle slightly roughly at 700 to 800 RPM (street cams) or 1000 to 1200 RPM for high-performance (off-road) cams.

3. If you want to lash the valves with the engine not running, follow Racer Brown's recommendations before you ever start the engine the first time. He suggests that the flywheel or crank pulley be marked to show the point of maximum lift for a valve, then rotate the crank exactly one turn right back to the same mark so that the tappet is in the center of the cam-lobe heel. Set the clearance on that valve, preferably with the P & G gapping tool. Proceed to the next valve and repeat the operation. Once the marks are permanently made on the flywheel or crank pulley and the order of setting the valves is established, this method is very quick and much more accurate than any other.

If it takes more than 10 minutes to do the first side, warm the engine by idling it again for 10 minutes with the rocker covers in place. The inspection procedures recommended in **2** should always be carried out, regardless of which valve-lashing method you choose.

Cylinder heads
all work better when seated and ported correctly

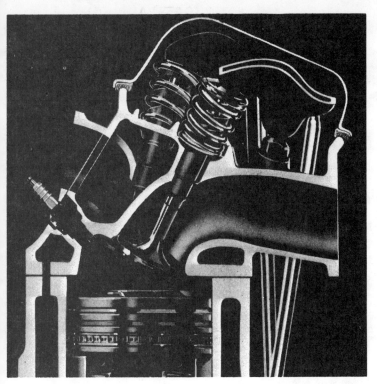

Transverse view of the cylinder head shows several distinctive features of the Turbo-Jet design: large smooth inlet and exhaust ports, valves set at unconventional angles, modified wedge-shaped combustion chamber and independent stamped-steel rocker arms.

HEAD DESIGN

In the introduction to the Society of Automotive Engineers' paper, "Chevrolet Turbo-Jet Engine," dated October 1965, the authors had this to say. "From the design standpoint, the most important characteristics of the new engines are superior breathing, increased combustion efficiency, and improved durability. To achieve the objective of high volumetric efficiency without compromise, the design and development program began with the establishment of the inlet port and exhaust port location and configuration.

"Once a cylinder-head design was established that provided the optimum in breathing, all other components were designed around this basic functional area.

"One difficulty encountered in designing V-8 engines is caused by the standard positioning of the camshaft between the cylinder banks. This location of the camshaft results in the pushrods extending up to the overhead area in such a manner so as to form a 'picket fence' through which the inlet ports have to be 'snaked.' This was not a problem with the Turbo-Jet engine. The ports came first and the valves were placed in a splayed position to accommodate the ports. Thus, the pushrods are placed far apart and do not interfere with the prime inlet-port locations. Actually, this may be an over-simplification since no portion of the cylinder head was designed without at least a slight compromise.

"Exhaust-port configuration has also been improved. The exhaust-valve position made it possible to design a larger radius exhaust port with a more gradual directional change and an unrestricted cross-section throughout the passage length. In addition, the offsetting of the exhaust valve toward the exhaust side of the engine minimized the port length, thus reducing the amount of high-temperature port area exposed to engine cooling.

"After locating all the ports, the valves, the valve springs, the rocker arms and the pushrods, there was very little room remaining to bolt the cylinder head assembly onto the cylinder block. As a result, the cylinder head is attached to the block with a bolting pattern unlike anything Chevrolet has had before. The new porting took preference and the bolts were placed to accommodate it. We have what we call a 6-5-6-5 bolting pattern with a long span in two areas.

"To fully compensate for the long span, a tapered bead is incorporated on the cylinder-head steel gasket. The taper amounts to 0.003-inch from one bolt to the next with the high point in the middle. Short reinforcing posts were cast into the head to stiffen this section so that proper gasket-bead compression would be achieved."

The inlet gas-velocity characteristics of the Turbo-Jet give approximately 16,000 feet per minute velocity for almost the entire length of the inlet passage. This is accomplished by keeping the areas in the inlet port in the head and in the intake manifold nearly constant throughout. The current interest in reducing emissions will see more and more effort devoted to keeping mixture velocities high as this can — with correct manifold design — reduce the need for running some cylinders 'way rich to compensate for those which are only getting a lean mixture. Over-rich mixtures are not wanted for low emission.

1971 low-HP 402 and 454 CID engines got this low-emissions head which is an open-chamber small-port type, part no. 3993818. Note that the non-plug side of chamber is straightened to a fore-and-aft line instead of angling from the intake towards the exhaust. This eliminated more of the quench area and thereby assisted in reduced emissions. A pair of these heads with a stout hydraulic camshaft and a spread-bore Holley carburetor would be ideal for a street screamer.

TYPES OF HEADS

If you were to look at the parts-book listings of cylinder heads for the big-block engines the chances are that you would be amazed at the number. The best way to get you squared away as to what's what is to list the heads and their major specifications. There are two basic combustion-chamber types: open-chamber and closed-chamber. The first open-chamber heads were introduced on the 2nd-design L-88's and ZL-1 engines for Corvettes. There are also two basic port types: large-port and small-port, with the small-port heads used on trucks and low-HP passenger cars. And, there are two basic types of metal used to make the cylinder heads: aluminum and Detroit's wonder metal—cast-iron.

Closed-chamber heads

This chamber is bathtub-shaped at an angle on a center line drawn between the two valves of each cylinder. It was the original chamber introduced on the Mark II Mystery Engines at Daytona in 1963. This chamber is small—only 108cc's, typically, and requires only a small lump atop the piston to create high compression ratios. However, it's considered "dirty" from an emissions standpoint. The closed-chamber design was used in a few applications in 1971, but was later phased out due to emission requirements. The open-chamber cylinder heads provide less emissions and are used on '72-and-later models.

These closed-chamber heads have been made in both large- and small-port versions and the large-port version has been made in both cast-iron and aluminum.

Intake valves in these heads are 2.06-inch diameter in the small-port heads and 2.19-inch diameter in the big-porters. Exhausts are 1.72 inches in all of the cast-iron closed-chamber heads except for the 1970-71 450 HP 454 engines which have 1.88-inch exhaust. Exhausts in the closed-chamber aluminum heads used on 1st-design L-88 engines are 1.84-inch diameter.

Open-chamber heads

By opening the chamber around the spark plug, Chevy engineers were able to get two improvements at the same blow: reduced emissions and improved breathing for higher HP. Reduced emissions are possible because the secondary quench area by the plug was eliminated. This really was not needed from a performance standpoint because quench is most useful farthest from the plug so as to get the mixture started toward the plug for more even burning. The first commercially available open-chamber heads were the 2nd-design L-88 and ZL-1 aluminum cylinder heads which became available in 1968. When combined with the camshaft provided with these engines, the end result was said to have 30% more flow capability than the previous cylinder heads. Part of this is due to port shaping and part due to the fact that the charge does not have to squeeze past the edge of the valve and around a combustion-chamber "wall." This makes the "semi-hemi" rat motor breathe even more

1971 454 CID cylinder head for LS-6 optioned Chevelle and Monte Carlo. The option was dropped prior to the availability of the head, but the head is available as a service part, 3994025. Chambers of this cast-iron open-chamber head are essentially identical to aluminum ZL-1/L-88 style heads used on 1969-71 Corvettes, including 1971 LS-6 in Corvettes. This head produces less emissions than previously used closed-chamber heads.

like the famous hemi. HP improvement obtained with the changes was obtained with less compression. The ratio was dropped from 12.5:1 to a 12:1 specification. The exhaust port was changed from square to round (shades of the Mark II!) and the valve diameter increased to 1.88 inches. Venturi-type seats are used for the exhaust valves, incidentally. On the intake side the contour of the port was changed to provide a slight venturi contour as the port approaches the valve guide. The pocket behind the valve has been filled in so that there is actually less port sectional area than previously. If you compare the ports on the aluminum open-chamber head with the ones used previously, you'll note that the port aims the charge at the backside of the intake valve. Some have noted that these ports look a lot like Formula 1 racing-engine ports with their venturi shapes.

Because the combustion chamber is larger—122 cc's—the pistons had to be redesigned for use with the open-chamber configuration. Even though a much larger dome is used on the pistons, the weight was kept the same so that the crankshaft counterbalancing would not have to be changed.

In 1971 Chevrolet made the open-chamber head with large ports available in cast iron. If you've priced the aluminum heads with all of the valves, springs and retainers, you know how many hundreds of hard-earned dollars they'll devour before you start any head cc'ing or porting.

Although Chevrolet made "no big thing" about it, they introduced small-port open-chamber cast-iron heads in late 1970 on some of the low-HP 454 engines. These have 2.06-inch intakes and 1.72-inch exhausts.

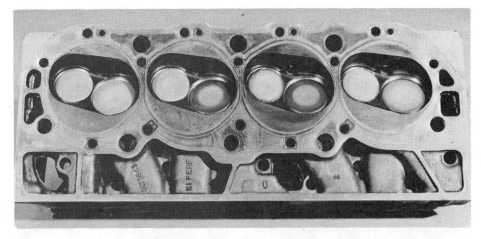

1970 450 HP 454 engines got this head with closed chambers, 2.19-inch intakes and 1.88-inch exhausts. Exhausts were 1.72-inch diameter in previous iron heads.

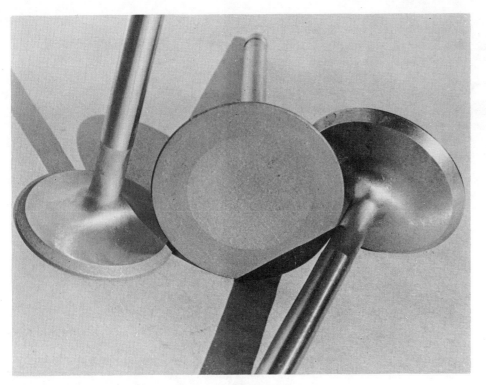

Stock Chevrolet 1.72-inch exhaust valve at left. Center is a 2.19-inch intake valves. At right is a 1.88-inch exhaust. Valves are shown approximately actual size.

VALVE SIZES

Chevrolet sells intake valves which are 2.06, 2.19 and 2.3-inch diameter. The 2.3-inch size is not used in any head which you can buy, but it is supplied to allow adding a larger valve if such is wanted. And, this valve can be used in a head which has been ported with too much enthusiasm or where the seats have been sunk excessively by incorrect seating or too many valve jobs. Kay Sissell uses the 2.3-inch intakes in some ZL-1 heads after trimming them to 2.25-inch diameter.

The 2.19-inch intakes can be added to any head equipped with 2.06-inch intakes and the 2.3-inchers can be installed in any head originally fitted with the 2.19-inch valves. However, we are not recommending the 2.3-inch valves, just indicating that they can be installed. The fact that these larger valves are not used in any of the Chevrolet heads could mean that they are larger than the head/port combination can use effectively.

Any of the cylinder heads can be fitted with the 1.88-inch exhausts, 3946077, to pick up another 10 to 15 HP.

1971 454 CID cylinder head for LS-5 optioned Chev, Monte Carlo and Corvette. This large-port cast-iron head is quite similar to all of the previous closed-chamber heads.

INTRODUCTION TO AIR FLOW

Porting

Porting is a term which has long been used to refer to modifications of the valve pockets behind the valves and the ports connecting the valve pockets to the intake and exhaust manifold flanges. For many years, engine modifiers "hogged out" the ports as large as they could make them, short of poking holes in the port walls. If the pocket behind the valve could be opened up to the inner dimension of the valve-seating surface, no one hesitated to do just that. Little attention was paid to the combustion chamber into which the valves opened. In rare instances, tuners noted that ported heads sometimes worsened the engine's low-RPM performance. But, even those that noticed it didn't stop making big holes—at least in their customers' heads—because that was what *sold*. It looked like the hot setup—whether it was or not.

As tuners began to recognize that flow benches could be used to measure the efficiency of ports with various valve sizes and lifts—with different port shapes—a lot of the old "hog-it-out" approaches were discovered to be detrimental. The use of flow benches has also caused hotrodders to become more conscious of intake-manifold-design requirements—including the size of the carburetor/s feeding the system.

Because many ported heads actually flow less air than the stock heads, we thought it highly important to provide the readers of this book with some guidelines as to what can be done to improve air flow in the cylinder heads without taking "giant steps backwards." *Porting* is now a "wrong" word because it does not embody all of the modifications which work together to affect air-flow characteristics through the intake and exhaust ports.

A lot of air-flow research is behind all of the big-block heads. Chevy's big-block engines have long proved their horsepower-making capabilities—boat racing is dominated by them. Because one of the important elements assisting the superb air-flow characteristics of these engines is valve shape, we used stock and modified Chevy high-performance valves in our air-flow tests. You'll see how important valve shape can be if you study cutaway drawings of famous racing engines, especially small-displacement ones. Such drawings reveal valve shapes which "work" with the port and valve pocket to get more air into and out of the engine.

Kay Sissell is one of several head modifiers in the country who insist that every port in the head be checked after reworking so that all ports of the same types will flow in a similar fashion. If a port is off, then further work will be required to get it to flow like its neighbors. Sissell says that exhaust-port work is the major requirement for the ZL-1 heads like the one he's flowing here. A manifold or injector is often attached to the head so that the flow measurements will include the entire inlet system.

Combustion-chamber shape

Combustion-chamber shape is also extremely important because the valves work into and out of the combustion chamber. A chamber shape which shrouds the valves so that they cannot work sharply reduces the air-flow capabilities. For this reason, you'll want to get open-chamber heads if you are seriously interested in winning—assuming class rules allow using these heads.

However, what shrouds a valve in one head may not be what shrouds a valve in another. Every engine and every head type is different. You can apply some of the things which are learned from one engine to another engine, but porting and combustion chamber design require careful investigation for each specific engine, head and cam combination on which you may be working.

Valve size becomes important because larger valves will usually enhance the flow characteristics of the cylinder head, even with poor port entry angles, especially when the chamber and pocket shaping are designed to work with the larger valves.

Flow bench tests

We originally planned to include details for seating valves and just give a passing reference to porting. But friend Kay Sissell—renowned six-cylinder expert in South El Monte, California—insisted that we use his flow bench to see what improvements could be made to the stock heads with stock valve sizes. As his own heads are known performance improvers for 6's of all types and large-block Chevrolets, Kay was hinting that there was good reason to perform these tests. This flow bench and Sissell's assistance provided significant details which should be interesting to all Chevy enthusiasts.

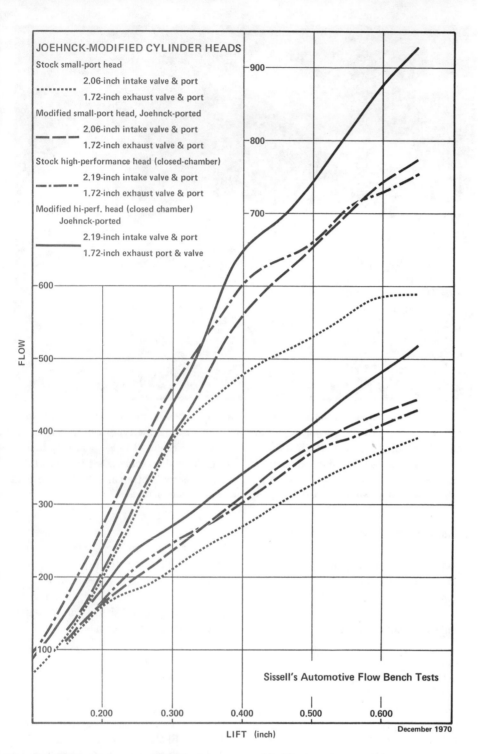

JOEHNCK-MODIFIED CYLINDER HEADS
Stock small-port head
  2.06-inch intake valve & port
  1.72-inch exhaust valve & port
Modified small-port head, Joehnck-ported
  2.06-inch intake valve & port
  1.72-inch exhaust valve & port
Stock high-performance head (closed-chamber)
  2.19-inch intake valve & port
  1.72-inch exhaust valve & port
Modified hi-perf. head (closed chamber) Joehnck-ported
  2.19-inch intake valve & port
  1.72-inch exhaust port & valve

Sissell's Automotive Flow Bench Tests
December 1970

INTAKE FLOW TESTS:
2.19-inch and 2.06-inch valve

Lift (inch)	1	2	3	4
.150	152	175	127	118
.200	240	270	208	200
.250	345	370	307	295
.300	440	460	392	393
.350	545	535	480	440
.400	650	605	560	480
.450	687	635	610	505
.500	745	660	655	532
.550	810	709	700	562
.600	875	730	742	587
.650	930	755	775	590

1. 2.19-inch valve in hi-perf closed-chamber iron head, Joehnck-modified port.
2. Same valve and head as in 1, in stock unmodified port.
3. 2.06-inch valve in small-port closed-chamber iron head, Joehnck-modified port.
4. Same valve and head as in 3, in stock unmodified port.

EXHAUST FLOW TESTS: 1.72-inch valve

Lift (inch)	1	2	3	4
.150	125	112	115	110
.200	186	168	165	162
.250	242	215	200	182
.300	271	244	238	212
.350	309	273	275	243
.400	344	365	312	270
.450	379	338	350	300
.500	412	365	380	328
.550	452	390	405	354
.600	484	410	427	372
.650	518	430	445	390

1. Stock swirl-polished exhaust valve with two angles under head to fair seat angle into underside of valve. Chamber edge of valve not radiused. Joehnck-modified port in hi-perf, closed-chamber iron head.
2. Same valve and head as in 1, in stock unmodified port.
3. Stock, non-swirl-polished valve with two angles under seat angle. Chamber edge of valve not radiused. Joehnck-modified port in small-port, closed-chamber iron head.
4. Same valve and head as in 3, in stock unmodified port.

Modified small-port heads really wake up the street-driven big-block engine. Don't "bad mouth" these heads because the ports happen to be smaller than the hi-performance heads offer. Chances are good that most street machines would be better performers if equipped with these heads or the new open-chambered cast-iron small-port heads described on page 78. This is especially true if the heads are ported as shown on page 86. 1.88-inch exhausts will help, too.

More about small port heads: page 110

Over 450 discrete flow measurements later we had found which changes would help to get more flow through stock-size valves. The flow bench showed beyond any doubt that the novice must confine his head-modifying activities to those explained in this section. Avoid "exotic" modifications until you have access to a flow bench, a pile of junk heads—and lots of time. Sissell's flow bench showed us that one of these can help to uncover hidden horsepower. Experts have long said that the engine is an ill-conceived air pump—if you can get more air into and out of the engine, more horsepower comes at the same time "for free." *Free horsepower is tough to get.*

We have been able to include actual flow diagrams of various big-block heads thanks to Kay Sissell and to Bob Joehnck who reworked both small- and large-port closed-chamber cast-iron heads. Joehnck also made us aware of the need for more radii on the exhaust valves than we had ever previously imagined.

The graphs show that the stock valves flow more as lift is added. Racer Brown had told us that flow-bench tests would quickly show why the big block "likes a lot of lift." Bear in mind that we only scratched the surface in this series of tests because the idea was not to write a whole book about head modifications. However, lest you think that this all comes quickly and easily, let me point out that we worked almost non-stop for one whole day on just the ZL-1 intake and exhaust and flowing the previously prepared Joehnck's heads. Let us assure you that we did not achieve perfection or maximum air flow on any port/valve combination.

From the curves you can see that the open-chamber head is the answer if you want real performance. Its box-stock intake flow is similar to that obtained after extensive modifications to the closed-chamber head. The exhaust is the area of greatest flow improvement. And, this is the area where most head modifiers spend their efforts on these heads—to make them flow still better.

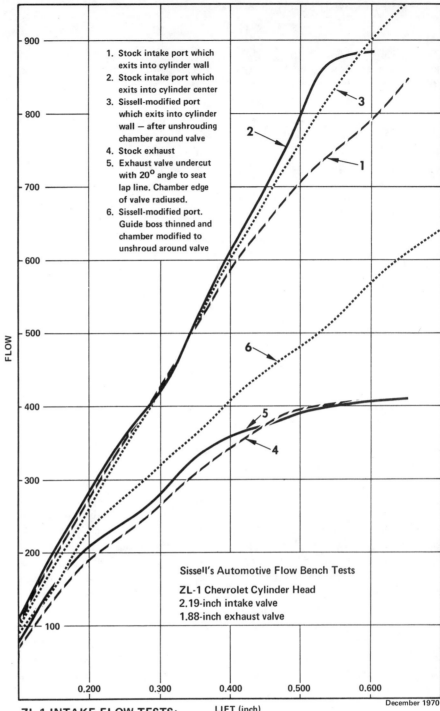

Sissell's Automotive Flow Bench Tests
ZL-1 Chevrolet Cylinder Head
2.19-inch intake valve
1.88-inch exhaust valve

December 1970

1. Stock intake port which exits into cylinder wall
2. Stock intake port which exits into cylinder center
3. Sissell-modified port which exits into cylinder wall — after unshrouding chamber around valve
4. Stock exhaust
5. Exhaust valve undercut with 20° angle to seat lap line. Chamber edge of valve radiused.
6. Sissell-modified port. Guide boss thinned and chamber modified to unshroud around valve

ZL-1 INTAKE FLOW TESTS:
2.19-inch valve

Lift (inch)	1	2	3
.150	192	200	180
.200	275	280	262
.250	355	360	345
.300	430	420	429
.350	510	517	512
.400	588	613	600
.450	647	695	681
.500	708	795	762
.550	750	875	835
.600	790	885	900
.650	845	—	950

1. Stock port which breathes into cylinder wall.
2. Stock port which breathes into cylinder center.
3. Sissell-modified chamber, with unshrouded area around valve, same port as Test 1.

NOTE: Test numbers are same as numbers on graph for intake port/valve only.

ZL-1 EXHAUST FLOW TESTS: 1.88-inch valve

Lift (inch)	1	2	3	4	5	6	7	8	9
.150	135	148	155	145	148	149	149	144	151
.200	194	208	210	208	207	209	208	210	232
.250	225	238	242	238	238	239	245	248	275
.300	268	280	282	275	283	295	300	301	320
.350	310	319	330	327	338	348	340	347	360
.400	345	350	360	370	384	388	388	389	409
.450	375	377	382	413	426	427	425	430	449
.500	398	390	392	467	470	470	474	468	482
.550	402	400	400	512	530	512	520	510	520
.600	408	408	408	555	570	555	565	550	570
.650	410	410	409	585	605	590	585	590	608

1. Stock port and swirl-polished stock valve; line 4 on graph.
2. Stock port, valve lapped to seat and 20° angle ground on back of valve to seat lap line.
3. Radiused chamber and seat o.d. on valve head edge; line 5 on graph.
4. Sissell-modified port; same valve as in 3.
5. As in 4, except back wall of valve pocket straightened on area nearest intake valve.
6. Guide boss narrowed on side next to intake. Note that this reduced flow except between 0.300 and 0.400-inch lift.
7. Other side of guide blended. This did not really help, either.
8. Tip of guide blended into port. This helped slightly from 0.200 to 0.450-inch lift. Conclusion: reworking guide is probably a waste of time.
9. Unshrouded valve with a radius into edge of combustion chamber; line 6 on graph.

NOTE: Another exhaust port in the same head was reworked without modifying the guide boss. It flowed 640 at 0.650-inch lift and had better flow throughout the lift range.

Exhaust Port

On the ZL-1 head with round ports, you can open the ports somewhat by the header flange. Chevrolet says that the next thing to do is to narrow the guide boss slightly, blending it into the port towards the header-flange side. However, as you'll see from our air-flow tests, this did not turn out to be all that helpful, so you may want to skip that step.

Don't make the common mistake of shortening the valve guide or cutting out the boss which supports it. Because of air-flow characteristics, more will be lost than gained. It's been proved on the flow bench, so don't doubt or argue!

Look at the accompanying photos to get a good idea of what's necessary.

Intake Ports

Blend in any sharp edges but don't try to enlarge the ports and don't polish the ports. Look closely at the accompanying photos and you'll have a good idea of what the professionals do when working on the heads. In general, getting the excess lumps out from casting irregularities and blending in the sharp edges are all that's possible for the neophyte engine builder.

Leave porting to the experts

If you are one of those die-hards that hasn't got the message yet—let's say that it is "smarter" to leave the ports and valves like the factory made them unless you have access to a flow bench so that you can see what removing a certain bit of metal from a port, combustion chamber or valve does to the air flow.

So when it comes to porting, get this job done by experts who use their flow benches and past experience to check that they are providing you with ports that will flow similarly. All exhausts should flow very close to each other, but there is no way to make the intake port that exits into the cylinder wall flow as much air as one that opens into the center of the cylinder. You'll be money ahead to buy the professional job instead of trying to do it yourself.

You might do a credible job of smoothing out some rough spots in the ports and possibly take away those sharp angles left when the head is machined at the factory. However, don't jump to the conclusion that you can do your own "basic" porting job and then have a professional finish the job to the ultimate for less money because of your efforts.

Those men rightly expect to start with a set of new and untouched heads and we can't blame them.

Prices vary according to your needs or wants. If you stop with the basic porting and seating job and keep the stock valves, then the job will probably cost a few hundred dollars. If a lot of special parts are purchased at the same time, such as titanium valves and retainers, etc., the price for reworking your heads could soar to more than a thousand bucks! Lest you think that the firms that do these jobs are getting rich at these prices, let us assure you that they are barely "making wages." There are a lot of hours of work in a good set of cylinder heads, far more than you'd ever guess unless you have done them yourself.

No high polish needed

You may wonder why there is no reference to polishing the ports when they are finished. The reason is that polished ports "look fast" but add no air flow. Each of the people with whom we talked pointed out that the old myth of highly polished ports is still widely believed. We thought that this had been buried years ago. So, let's try again. The finish on the ports, so long as the bigger lumps have been smoothed off, does not require polishing. In fact, the slight roughness left by rotary files or grinding stones can actually be beneficial in promoting vaporization of any puddled fuel in the ports.

Polishing ports takes an enormous amount of time, is an especially dirty job on either aluminum or cast-iron and consumes expensive sanding sleeves to get the final slick finish. And, if such has been done, what do you have for this unnecessary investment of time and materials? Eyewash! Nothing else.

Jim Caravello of Diamond-Elkins Porting, Bill King of King Engine Service and Paul Hogge of Racing Parts and Machine all took special care to say that polished-head hang-up is so bad that heads are occasionally sent back because the customer thinks that the heads could not possibly be any good unless he sees himself reflected in the port surfaces. If you find someone who insists on polishing the ports in your heads, you'll probably be wasting your money with an old timer who has never checked romance against reality on either a flow bench or a dynamometer. So, if you *must* polish something, use your energy that might have been wasted on the ports to polish the combustion chambers very carefully without changing their shape. Polishing the valve heads and combustion-chamber surfaces, as well as the piston tops, reduces the heat loss to these components, thereby improving power output. Polishing provides a secondary benefit by making carbon removal easier when the engine is taken apart for cleaning or a valve job. Sanding sleeves and Cratex abrasive-impregnated rubber polishing tools work fine for this chore, but use them in your electric drill. If you own a high-speed air or electric grinder, do not run any tool above its published RPM limits. If the tools are sold from an open bin, or with no instructions—find out what safe speeds are in the manufacturers' literature or you can injure yourself when a tool bends, breaks or disintegrates. It may take extra effort or even a letter to get the manufacturer's data sheet, but do it! Wear goggles or safety glasses every time you do any cutting or polishing. Chips in the eye are painful, time-consuming and expensive—and painful.

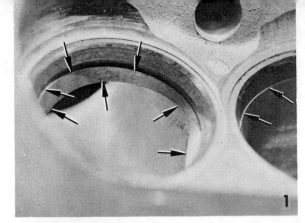

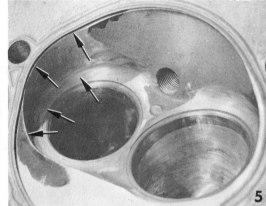

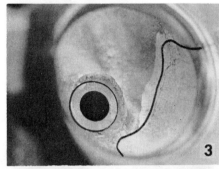

These photos are of the Sissell-modified ZL-1 head depicted in the flow graphs and tables on pages 83 and 84.

1, 2 — Exhaust port, combustion chamber and part of intake port prior to unshrouding and before port work. Arrows indicate sharp edges which will be radiused and blended for best flow.

3, 4 — Exhaust port after radiusing the entry under seat into port runner. Black line indicates area blended into port pocket. Guide boss has been thinned slightly on one side. This is test 5. 4 shows a slightly different view of the exhaust port where underside of valve seat has been gently radiused into port runner. Note that the venturi shape under the seat into the port has not been destroyed.

5 — Exhaust has been unshrouded by opening the edge into a scribed line indicating gasket edge. Arrows indicate extent of area which has been reworked. 25° top angle on exhaust and 35° top angle on intake are completely blended into the chamber. Note that the 70° undercut has been blended into intake port and intake valve has also been unshrouded. All of the machined cuts have been blended into the combustion chamber. Chamber still requires smoothing with sanding sleeves and edges of chamber must still be radiused.

6 — Intake after blending. Guide does not have to be removed or driven back out of the way. Note that a slight lip or venturi has been left just ahead of the seat where the port enters the backside of the seat. Photo 5 shows intake port from another angle.

Photos of Joehnck-modified heads depicted in flow graphs/tables on page 82. Small-port above, big-port below. Pay special attention to gentle curves used throughout. Note that pockets under valve seats do not dive straight into the runners. There is a gentle lip or radius under each seat. Each valve is carefully unshrouded and seated with the three angle factory-recommended method. Note that the exhaust guide bosses are merely smoothed. Don't remove 'em!

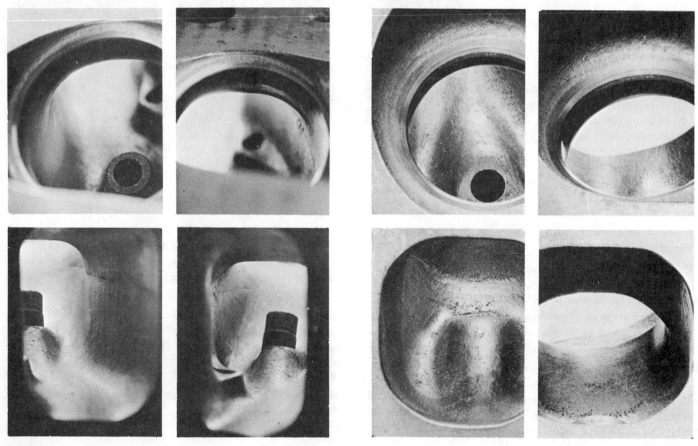

VALVE SEATING

Whether you start with the open-chamber head or stick with the closed-chamber version, you're probably burning with desire to do something to your heads. Chevrolet's recommendations for seating the open-chamber ZL-1 heads follow, but be assured that the plan works for any head, whether open- or closed-chamber and whether yours are aluminum or cast-iron.

Close attention to the valve-seating task adds HP with such a minimal investment that you could almost call it "free." It's the factory's gift to you—just a little loving care and attention make it happen.

Stock valves overhang the seat opening by a large amount, reducing flow capability of the valve/port combination and stealing *at least 10 to 12 HP!* You need to trot out *all* of the horses from your engine's stable because your competitors will get them out—believe that! But, understand that there is a lot of metal to be removed to do the seat job recommended by Chevrolet. Most high-performance-oriented machine shops will charge about $150 for that first competition valve job, and it is no money-maker as far as they're concerned. It is especially tough doing the first good valve job on the aluminum heads because the steel seats are harder than the hinges of Hell. However, every big-block engine needs this type of valve job to get maximum flow capability, even a brand-new one.

You'll note that it is so simple that any competent machinist should be able to do it—if he has the time and doesn't run out of patience because you've hammered him so far down on price that he can't make wages.

Make sure that the seats are checked with a seat indicator gage to ensure that seat runout does not exceed 0.001 inch. If the machinist does not have this kind of measuring equipment, don't take his word for it that he's measured his tools and they always produce this kind of accuracy. Baloney! Each and every seat has to be checked with the indicator to see whether that kind of accuracy has indeed been achieved *on that seat*. There's no way that this can be done unless the measuring tool is right there. What we are saying is that it is possible to have beautiful-looking seats which are 'way out

FACTORY VALVE-SEAT RECOMMENDATIONS

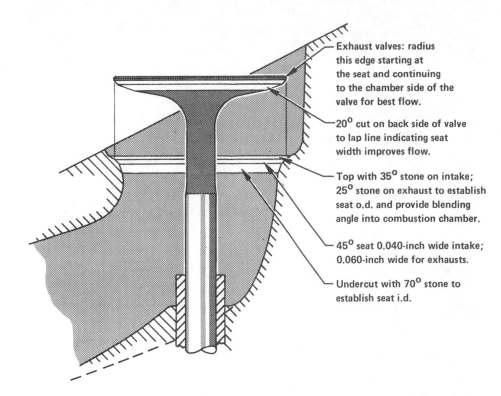

There's 12 to 15 HP just waiting for you if you'll just do a seat job according to the factory's recommendations — as presented in the drawing above. But, it's not a job for a vacuum-cupped stick and a can of valve-grinding compound. This requires four to five hours — or more — with professional valve seating and grinding tools.

and will not last. To get the best flow through your new seats, we have also added additional details of radiusing which can be carefully done with a hand grinder, fine stones, sanding sleeves and Cratex tools. The inside and outside of the seat can be radiused into the top and bottom angles.

Lapping the valve against the seat won't tell you anything except where the seat is in relation to the valve. Lapping does show you where you can take the back angle cut to, however.

The circumference of the valve seats should be increased to equal the valve o.d.'s. Then the inner diameter of the inlet seat can be ground with a 70-degree stone (we'll talk about fancy multi-angle valve jobs later on) to get a seat which is at least 0.040-inch wide. Use the same technique to get 0.060-inch-width for the exhaust seats. The valve seats themselves are all ground at 45 degrees. A further topping cut above the seat can be made with a 35-degree stone on the intake and a 25-degree stone for the exhausts. The edges of these cuts should be blended into the combustion chamber with rotary files, mounted grinding stones and subsequently with sanding sleeves and Cratex tools. The edge of the topping cuts can be blended into the 45-degree seat with a slight radius *if you have a steady hand* and if you have an adequate supply of fine-grit sanding sleeves and Cratex at hand. The 70 degrees can be blended into the 45-degree seat similarly. And, the 70 degrees can be blended into the port throat with rotary files or mounted grinding stones.

Further aid to flow is given by making a 20-degree cut on the underhead of the valve itself so that the valve face is matched in width with the valve seat. The edge of this cut can be radiused into the 45-degree seating portion and the area at the other end of the 20-degree cut can be blended further into the backside of the valve head with additional blending cuts which you can "eyeball-in" with the valve grinder.

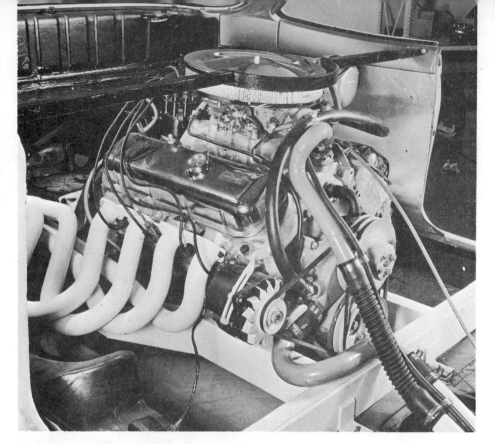

Strange home for a big block: the back seat of a Corvair 65-69 coupe. Although the transaxle likes the torque of a small block much better, big blocks have been used in the very well engineered Corv-8 conversions engineered by Crown Manufacturing's Ted Trevor. Even with a small block, the performance is very Group 7-ish!

If you are really serious about racing your big block — or any other engine, for that matter — get a dynomometer and find out what's really happening. Compact units bolt directly to the bellhousing in about 10 minutes. Simple stand as shown here works fine. The cost might seem a bit high, but you'll save that much very quickly — and start winning races with more HP — by using one.

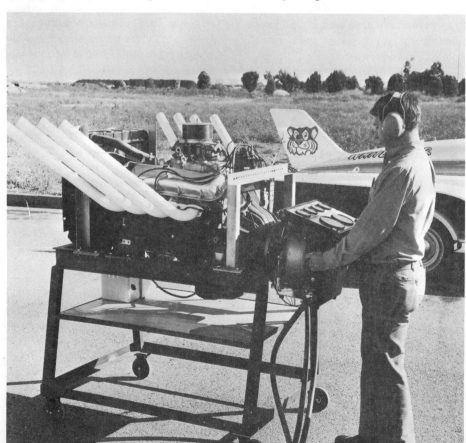

Wider seats may last longer

Although narrow seats have long been considered optimum for racing, especially when the seat edges are radiused into the adjoining angles, not all experts agree on this. And, it is well known that narrow seats must be right on as regards run out — within 0.001 inch, certainly — if the seats are to live very long. Our flow-bench tests to date have not shown any hindrance to flow from using wider seats. In fact, this sometimes improved flow over narrower seats. Bob Joehnck suggests that the intake-seat width be 0.100 inch at least and that the exhausts be 0.070 to 0.080 inch, even though the seat on the valve itself may be narrowed on its inside diameter with a $20°$ cut — or multiple angles — to promote flow. More research is probably needed to determine the optimum seat width, but the wider seats should last longer for a street-driven machine.

Radius Seats & Multi-Angle Seats

Flow-bench work has shown that more air will flow through a valve/seat combination if there is a radius on the seat instead of a flat surface formed by a discrete angle. Similarly, the valve itself can be faced with an opposite radius so that the two radii are tangent to each other when the valve is on its seat. In this instance, the actual line of contact may only be 0.020 to 0.030 inch, as established by lapping to show the contact area. Seat runout must be tightly controlled with the radius-seat setup.

Radius seats can be machined into the heads with cutters or ground in with radiused stones. The valve can be made to approach a radius configuration by grinding a series of angles onto it and then blending these into a radius with a specially set up valve grinder.

An approximation of the radius setup can be accomplished on the cylinder head by using a series of angles. This has been done for many years by knowledgeable racing mechanics.

We should mention that such work is really only for the all-out racer and is not for the street-driven machine.

N.H.R.A. HEADS

Getting your "stock" heads into an N.H.R.A.-legal state is worth a few hundred dollars, regardless of who does it. There's cc'ing, surfacing of the heads and the detailed seat job which requires only three angles on the seat with the angle under the seat extending into the port a maximum of 1/4-inch below the seat i.d. The valve can have only one angle.

Paul Hogge, having fought his share of verbal battles with "drag-strip lawyers," insists that his customers get a sheet from the division N.H.R.A. technical director with the specifications to which the engine must comply. That way, the competitor has the information on an N.H.R.A. letterhead with a signature and a date to back up the specs to which he has built and checked his heads and other engine components.

Modified classes, on the other hand, can be run with multiple angles on the backside of the valve and multiple angles on the seat/port combination. Even the expensive-to-do radius seat job which blends a radius from the port right into the chamber can be used in modified classes.

COMBUSTION CHAMBERS

Next the combustion chamber is tackled to remove any ridges or sharp spots, especially at the edge of the chamber and around the spark-plug entry. Again, use your hand grinder or 1/4-inch drill with suitable devices mounted in its chuck. Most builders will blue the cylinder head with layout blue, then scribe each chamber to show the gasket opening. Some valve unshrouding and blending of the chamber wall to equal the scribed lines is common practice. And, for the exhausts, especially, can be extremely productive in terms of increased air flow.

Screw a spark plug of the correct reach into each of the chambers and look at how the end of the plug mates with the chamber. You may have to take threads out of the chamber. If the plug should protrude into the chamber, see the chapter on ignition for details of thick plug washers. You do not want any exposed threads on the plug or in the head itself.

Do not try to "open-chamber" your closed-chamber heads. Refer to the accompanying photo which was made of a closed chamber head. Kay Sissell sawed this one through the center of the plug area so that his customers could see for themselves that there is not enough metal in one of these heads to allow laying back the plug side of the head by more than about 1/4 inch *maximum*. We know that various head modifiers have advertised that they can open-chamber your heads. Maybe they *can?* Study the photo and then decide whether it's worth the risk, especially now that there are both small- and big-port cast-iron heads available in the open-chamber configuration for very low $$.

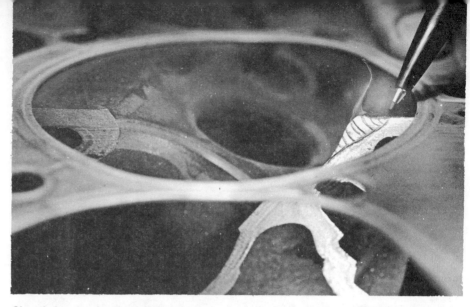

Closed-chamber head sectioned through plug hole shows that laying back the chamber into a full open-chamber configuration would get into the water jacket. Kay Sissell has marked in the area which could be laid back for better breathing. It is about half the distance between the edge of the stock chamber and the gasket. Even this could cause some trouble in a head with casting flaws. Although compression is reduced by this modification, HP goes up.

On some competition-only large-block Chevys, it is necessary to machine a groove around the combustion chamber and then press in a small, soft copper wire to prevent combustion leakage. This irregularly shaped groove is machined in a vertical mill with a tracer attachment. This is definitely not needed for street or even mild race-car applications, but is shown here to familiarize the reader with the much talked about process of O-ringing.

VALVE GUIDES

Here is one of the areas where Chevy failed to make it easy for big-block mechanics. They don't sell replacement guides for either cast-iron or aluminum heads—but they will sell you a new head, of course. Here's why: the factory installs semi-finished guides in the heads and finish bores them when they machine the seat and port. Thus, the guide outside diameter is seldom concentric with the seat, but the guide i.d. is right on. Thus, if you install a new guide with inside bore and outside diameter concentric—which is what you would expect—these guides will typically be 0.020 inch out-of-concentric with the seats. Correcting this condition always requires sinking the valves so much that you gain a lot of cc's in the chamber, thereby lowering the compression and requiring that you go through the cc-ing routine again. So, perhaps you can see why Chevrolet does not sell replacement guides. And, this is why we recommend that you use the bronze-wall guide-rebuilding scheme to save the time and money that you'd otherwise spend in getting those seats redone and the chambers re-cc'ed.

Here is one of the areas where Chevy failed to make it easy for big-block mechanics. They don't sell replacement guides for cast-iron heads—but they will sell you a new head, of course. Here's why: the factory installs semi-finished guides in the heads and finish-bores them when they machine the seat and port. Thus, the guide outside diameter is seldom concentric with the seat, but the guide i.d. is right on. Thus, if you install a new guide with inside bore and outside diameter concentric—which is what you would expect—these guides will typically be 0.020 inch out-of-concentric with the seats. Correcting this condition always requires sinking the valves so much that you gain a lot of cc's in the chamber, thereby lowering the compression and requiring that you go through the cc-ing routine again. So, perhaps you can see why Chevrolet does not sell replacement guides. And, this is why we recommend that you use the bronze-wall guide-rebuilding scheme to save the time and money that you'd otherwise spend in getting those seats redone and the chambers re-cc'ed.

1968 Corvette L-88 1st design heads, bottom and 1969 L-88 2nd design and ZL-1 heads, top. Late heads have open combustion chambers for improved breathing and efficiency. These heads have large round exhaust ports, venturi-type exhaust seats and streamlined intake ports for improved flow characteristics. Exhaust valves in the open-chamber heads are 1.88-inch diameter. Note the differences between the two heads, especially in the valve-guide areas.

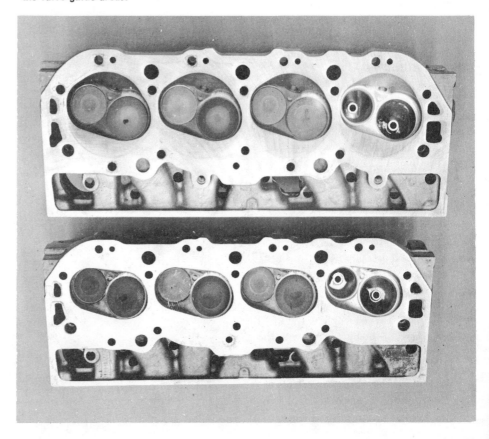

Valve-Guide Inserts—At the top of my list of valve-guide fixes is the guide insert. Guide inserts can actually restore guides to better than original. Guide inserts come in two different styles and several materials. The first type is a sleeve that is driven in, the second is threaded in and looks like a Heli-Coil.

Sleeve-type inserts are available in cast-iron, bronze—sometimes called *phosphor-bronze*—and silicon-bronze. All require the original guide to be reamed oversize so it will accept the insert and so there will be an interference fit between the guide and the insert. After reaming, the insert is driven into place. Some sleeve inserts are presized, while others need to be reamed or honed after they are installed.

Sleeve-type cast-iron inserts have the same wear characteristics as the original guides and are reasonably priced. Bronze inserts wear better than cast-iron inserts, so it follows that they cost somewhat more. Silicon-bronze inserts have superior wear characteristics; 150,000 miles is normal for a set of these, but they represent another step up in price.

Sunnen makes a kit that uses silicon-bronze inserts and requires honing them to size after installation. Some machinists say these are the "cat's meow" because of the smooth finish left on the bronze insert by the hone. The price compares favorably with other reaming and installation procedures and the smooth finish provides maximum thrust area. This translates into long guide and stem life.

Thread-in inserts first require tapping the guides. The tap used must be sharp so that the threads are smooth. If the threads are not smooth, the insert will not fit tightly in the guide, allowing the insert to move up and down with the valve. This movement pumps oil into the port, so you've got to be careful when installing these inserts. Additionally, a well-cut thread ensures that insert-to-guide contact is good to allow for maxmimum heat transfer from the valve to the insert, to the guide and, then to the head. If the insert doesn't transfer heat properly, the valve stem will overheat and expand, and you could get a sticky valve.

After threading it in, the insert is locked into place by expanding it with a special tool.

New Guides—If you absolutely insist on installing new guides, here are some points to remember:

The exhaust guides are stepped slightly. They must always be driven out from the port side so that the guide exits from the head on the rocker side. Similarly, guides must always be installed from the rocker side and driven in toward the valve side. Measure the amount that the guides extend out on the rocker side before taking them out. Then, be sure to reinstall the guides with the correct amount protruding from the head.

Remember that the exhaust guides are exposed to water. Sealing is essential at both ends of the guide to avoid problems, so coat the guide with anti-seize before driving it into place. It won't hurt to coat both openings in the head, too. The anti-seize will act as a lubricant and a sealant—helping to ease the installation and then ensuring that the finished job won't leak. A hydraulic press could be used to install the guide, but this would require a multi-angled fixture to support the head rigidly and correctly or you'd wreck both the guide and the head.

You have several choices if you are forced to buy new guides. Easiest to purchase are cast-iron guides made by Safeguard or Manley.

Another way to go is to use silicone-aluminum-bronze guides. These are popular on air-cooled engines such as Porsche, VW and motorcycles. However, because these work against enormous spring pressures and rocking due to high lifts, spend the time required to get the push-rod length correct so that the valve-train geometry is exactly right—to avoid the even more time-consuming problems and expense involved in guide reworking.

Another of Chevrolet's best ideas: slip-on valve-stem seals. Plastic capsule on end of installed valve stem is lightly oiled so seal slides past keeper groove without damage, is removed after seal is on stem. Installer is also part of seal package 3998265. Seals are last items to install before the springs because pulling a valve back through the seal will destroy the seal lip.

VALVE STEM OIL SEALS

Oil seals can be used to reduce oil entering the engine through the guides. Stock ones are rubber umbrellas which fit over the valve stem and guide, or nylon umbrellas built into the retainer caps. Because the latest steel retainers do not have built-in seals and won't work with rubber umbrellas, Chevrolet has supplied a Seal Unit. This kit includes eight seals which install right onto your guides *with no machining required!* A plastic installer is slipped over the installed valve stem and a drop of oil is placed on the plastic so that the seal will slide over the stem and past the keeper groove without any damage. Then a steel installing tool is piloted over the stem and used to drive the seal onto the guide *gently* until the seal bottoms on the guide. Further blows on the installer after the seal seats will distort the seal and make it useless.

All of Chevy's springs will install over these seals with adequate clearance. Seals are available from other manufacturers, of course. But, when you stop to consider that these all require machining the guides, why bother?

HARDWARE
Head-bolt washers

Hardened steel washers for use on the aluminum heads are available as a stock Chevy part or from JEB Industries. If you use these, you must make sure that there is sufficient thread engagement in the block. You may prefer to use the longer bolts designed for the aluminum heads (see the Heavy Duty Parts List) with washers on your cast-iron heads, again making sure that there are enough threads on the bolts.

Studs

Many professional racing engine builders use studs in the cylinder block and torque the heads into place with nuts. Chevrolet has now released a stud kit just for this application, P/N 3965763. These studs should be used with hardened washers P/N 3899696 and nuts P/N 3942410. Cost for the stud kit — less the washers and nuts — is under 40 dollars. JEB Industries also supplies high-quality studs, hardened washers and aircraft-quality machined nuts. A set for your big-block heads will cost about $65.

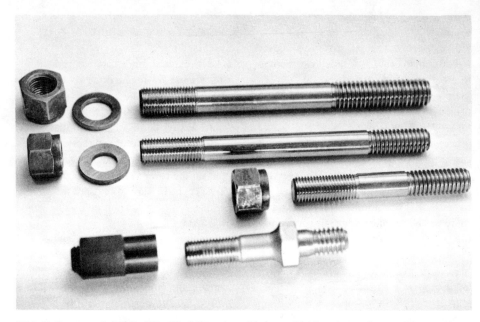

Dick Belleveau of J.E.B. (Air World) makes this beautiful hardware for professional rat-motor builders. Top is main stud, made in $50 sets for both iron and aluminum blocks. Center items are cylinder-head studs which are $65 per kit. These are also made for aluminum and cast-iron blocks and heads. $2 rocker stud (bottom) is squared across tip to allow correct wedging action with socket-head set screw in positive-locking adjusting nut. You won't find better hardware than this.

COMPRESSION RATIO

In general, the average dual-purpose big-block, or even an engine in a car used solely for transportation, should be equipped with a true compression ratio of 9:1 to 9.5:1 maximum. Premium gasoline is required with this compression ratio.

Racing engines may be able to tolerate compression ratios of up to 11.5:1, but anything this high will not run very well on any gasoline that you can buy in a service station . . . and the problem will worsen with the swing to non-leaded gas. The reason that a racing engine can tolerate a higher compression ratio is that camshafts with lots of overlap reduce the engine's low-speed pumping capabilities. Higher compression ratios offset some of the loss of low-speed torque caused by large amounts of cam overlap.

Tom Langdon, a Chevy high-performance engineer interviewed by Motor Trend in 1969, said, "We seem to gain horsepower on the dyno up to around 11.5 or 11.75:1 c.r. Over that point the dyno shows no change. I would say that in the field it's very possible you would see a loss."

QUICK COMPRESSION RATIO CHECK

To measure the approximate compression ratio of any engine requires only the burette described in the following section. Bring a piston to TDC, seal the top piston ring to the cylinder wall with a quick smear of grease around the edge and install a head gasket of the type to be used on the engine. A used one is fine. Install the head and tighten the bolts. Turn the engine so that the spark plug is vertical (easier if the engine is out of the car!). Remove the plug and fill the cylinder with the same fluid used for head CCing. Note how many cc's this required to get to the bottom of the plug threads. Convert the chamber measurement to cubic inches by dividing the cc's by 16.386.

$$\frac{\text{chamber displacement} + \text{cyl. disp.}}{\text{chamber disp.}} = \text{c.r.}$$

Example: 4.094-inch bore x 3.76-inch stroke 396 has a cylinder displacement of 4.094 x 4.094 x 0.7854 x 3.76 = 49.493 cubic inches. Measured chamber capacity 111.2 cc's ÷ 16.386 = 6.786 cu. in.

$$\frac{49.493 + 6.786}{6.786} = 8.29:1 \text{ c.r.}$$

Note: The plug hole is seldom the highest point in the chamber, so there will nearly always be some air left in the chamber. Therefore, calculations based on this method typically give a higher compression ratio than you actually have in the engine. Your calculations will be most accurate with an open-chamber head.

Open-chamber aluminum head combustion chambers are quite large as compared to the closed-chamber heads. This shows you why it is important to use the correct piston with these heads. Otherwise, the large chamber volume would give you very low compression. Numbers written on head are guide diameters. Minor grinding in the ports and elsewhere on the head will often be evident on a new head straight from Chevrolet. The chambers on these heads must be deburred when you get them, especially around the edges and where the plug hole threads enter the chamber.

CYLINDER HEAD CCing

Generally, the assumption is made that in a modern multi-cylinder engine all cylinders are doing an equal amount of work and contributing their fair share to the total horsepower and torque output of the engine. This is not the case. For the cylinders to produce equal amounts of work, the total volume of each must be equal. As the engine comes from the factory, they are nearly equal—but there's a not-so-obvious inequality in the big-block Chevrolet. All of *its cylinders do not operate the same* because of differences in the combustion-chamber turbulence. So, when you make all of the cylinders breathe in the same volume, you actually have two four-cylinder engines, each with different capabilities because of turbulence differences. The highest turbulence chambers, 2, 3, 6 and 7, are the ones in which Chevy recommends using less spark advance via retracted-gap plugs. The difference in turbulence is due, of course, to the angle at which the port dumps the fuel charge into the cylinder. According to Chevy Engineer Bill Howell, it's actually possible to run more compression on the high-turbulence chambers *if the correct spark advance is used* to take best advantage of the turbulence. However, he carefully pointed out to us that these are the chambers which will detonate first, so real care is needed to keep the advance and mixture correct in them.

Now that we know that the cylinders do not all work alike, why should you spend the time and/or money to get your heads cc'ed to the minimum allowed for the class in which you will be competing? Well, this does produce the highest compression which is allowable. And, it does so for all of the cylinders. Because of the way the cylinder heads are made it would be difficult to raise the compression to the legal limit in only the high-turbulence cylinders. Furthermore, the lower turbulence cylinders can utilize all of the compression that is allowable in any stock class so it is still important to get *all* of the chambers nearly alike at the allowable volume. The less obvious, but just as important second reason is the need to know the compression ratio so that it can be adjusted if it is too high or too low for the intended use of the engine, fuel octane, etc.

The process of making all of the cylinders nearly equal in volume is called CCing. And, this provides the measurements which you need to calculate compression ratio. Getting the volumes equal not only requires making all of the combustion chambers equal, it also requires equal deck clearance at each combustion chamber.

If you are preparing a set of heads to be used in a competition engine which must "pass" as legal if torn down for technical inspection, be sure that you read and understand the rules. Know what is required for minimum head volume (hence the need for CCing as we are about to discuss), what chamber modifications are allowable—if any—and what type of valve-seating job is permissable for the class.

The job is not difficult, but is time-consuming and requires considerable patience on the part of a novice. A relatively small amount of equipment and money is involved, except for the final steps which require the services of a competent automotive machinist.

Begin with a cylinder head which is spotless. If the head has been run before, have it sandblasted, glassblasted, boiled out or scrubbed clean with a stiff wire brush attached to the end of an electric drill. It is most important to remove all deposits of carbon from the combustion chambers. Make certain that the machined surface of the head which mates to the block is straight. This may be checked in most cases with a steel straight-edge such as a large machinist's rule. If there are any doubts in this area, have a

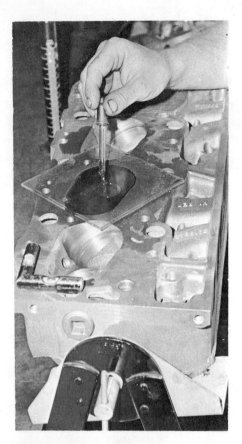

Head CCing can be done with an eyedropper but to save your sanity—buy a burette. Here the dropper is used for the final filling.

machine shop mill or grind the head 0.003 inch to true the surface. Rough in a valve job and install the valves, springs and keepers and the type of spark plug to be run in the engine. Roughing in the valve job means having a valve job done on the head up to the point of making the final pass with the finishing stone. This will be done after CCing has been completed. As the valves are being installed in the head, coat the valve seat with a very thin coating of lightweight grease. You want a liquid-tight seal between the valve and the seat. Because the valve job is not yet finished, we cannot rely on the two metal surfaces to accomplish this seal—hence the light grease coating to retain liquid.

For the actual CCing, you'll need a chemist's burette and a small piece of plastic. This plastic should be a minimum of 1/2-inch thick and large enough to completely cover one combustion chamber with at least two inches of overlap all around. Drill a small hole near one edge of the plate. The hole should be about 1/4-inch in diameter and be countersunk on one side with a larger drill in order to ease filling the chamber.

The other tool needed for the job is the chemist's burette. This instrument is nothing more than a graduated glass tube with a petcock at one end to dispense accurately measured amounts of liquid.

For a measuring liquid you won't have to look far. Mix automatic transmission oil and clean solvent in about equal amounts. You'll probably need about a quart for each head. The color and viscosity of this mixture contributes greatly to the ease of measuring.

Block the cylinder head up on a bench or table so that one side of the head is higher than the other by about an inch. This is done so the hole in the plastic plate may be positioned at the upper edge of the chamber and the combustion chamber can be completely filled with the measuring fluid. If the machined surface of the head is set level, then it becomes all but impossible to chase the bubbles out of the fluid when attempting to determine accurately how much fluid is contained in the enclosure. The head should be clamped or blocked so that it will not move during the operation. Fill the burette with Ol' Red (the measuring liquid) to the zero marking. Use the petcock to drain some of the liquid out to get the level exactly on the zero mark. This is critical! Note that there are two types of burette markings: zero at the full mark and the capacity of the device at the full mark, i.e., 100 cc's, etc. If you are buying a burette, get one that is marked with zero at the full mark because it reads directly as the fluid is metered out into the head.

Before pressing the plastic block into position over the chamber to be checked, coat the outside edges of the chamber with a very light coat of grease. Spread the grease evenly; then press the plastic block into position, leaving the small hole near the upper edge of the combustion chamber. With the burette over the hole in the plastic plate, slowly open the petcock and begin filling the chamber. As the fluid reaches the seal between the plastic and metal watch for leaks. If there's any seepage between the head and the plate, you'll have to go back to the start. Remove the plastic plate, pour the fluid out of the chamber, reseat the plate and remeasure the liquid. Don't attempt any shortcuts, 'cause there ain't any.

Paul Hogge of Racing Parts and Machine equips his plastic plates with head-bolt holes so that bolts can be used to clamp the plate to the head without any grease for sealing the plate so that no error will be introduced by a heavy coating of grease.

Barring complications with the seal, fill the chamber until the liquid just touches the bottom of the hole in the plastic plate. Although it may seem nit-picking, check the chamber for bubbles because they displace volume and will affect the measurement of the chamber. Carefully record the amount of liquid in the chamber; move on to the next chamber and repeat the measuring process.

We'll now assume that all chambers of one head have been measured and we have readings of 109.8, 110.2, 111.2, and 109.0 cc's (cubic centimeters). This may seem close enough—and it is for the run-of-the-mill engine. For our purposes we're going to make all the chambers measure the same. To aid our explanation process here, we shall assume that the head is going to be used on an engine in some stock class of competition which will not allow any removal of metal or polishing inside the combustion chamber. Let us also assume that rules from the same racing association will not allow the addition to or removal of metal from the face of any valve. As a further stumbling block, let us assume that for this particular engine the minimum head volume per cylinder is set at 109.8 cc's.

The next step is to bring all chambers up to the volume of the largest of the group—in this case, 111.2 cc's. This is accomplished by "sinking" valves. This nomenclature describes moving the valve seat further down into the head to increase the volume of the chamber. Careful thought should be given before any metal is moved around. It is most important to remember that any valve which is sunk cannot be raised. Keep planted firmly in mind that overly enthusiastic activity in the area of valve sinking will soon put you in the market for a new cylinder head.

Any valve which is sunk changes the flow characteristics in the cylinder and port area. An intake valve should not be sunk unless it is absolutely necessary, and then not more than 0.040 inch. This should be considered the absolute outside limit. Sinking an exhaust valve is not nearly so critical, so if you must sink a valve, make it an exhaust.

The process of sinking a valve runs something like this: Top an intake valve seat with a 35-degree stone; an exhaust with a 25-degree stone. This unshrouds the circumference of the seat. Next, re-establish the valve seat with a 60- or 70-degree stone, then a final pass with a 45-degree stone. Now carefully clean the chamber, install the valves, springs, retainers and spark plug and follow the previous steps until the chamber will hold the exact amount as the largest chamber in the head (in this case—111.2cc's).

Although it might appear that there is no simple method to determine how deep the valve should be sunk to gain the needed volume, it just takes a simple formula. The valve area times the amount sunk equals the volume which will be gained. However, the whole deal is painstakingly slow going. The first set of heads will be the hardest—on this set you'll want to take off only the slightest amount of metal on each pass. After running through this process on a couple of cylinders, you'll most likely determine that one is close enough.

Remember that this should be only when all chambers measure *exactly* alike. What do you have to lose if one cylinder runs a cc off? You tell us when the other guy wins the race.

With all chambers now measuring the same, the valve job may be finished off. With repeated passes of the stone to sink the valves, some of the seats might be quite wide. The remedy is to narrow the seat with a 55-degree stone and then move into the throat of the port for a pass with a 70-degree stone (if four angles are allowed by the class rules).

To round out the process of head CCing we need to reduce all the chambers to 109.8 cc's. As you recall, all of them now measure 111.2 cc's which is too large to provide the allowable compression ratio for the class.

With valves and spark plug installed in any of the chambers, block up the cylinder head until the machined block mating surface is exactly level. Be sure that you are using the type of spark plug which you plan to use in the engine. *The head must be level.* Fill the burette and then carefully drain exactly 109.8 cc's into the chamber. Rig a depth micrometer over the chamber to indicate the difference between the machined, level surface of the head and the surface of the fluid. Read this micrometer just as the fluid in the chamber begins to "jump" from the chamber to the micrometer shank as it is lowered toward the fluid. The reading at this point is the amount of metal to be ground or milled from the head to produce the desired chamber volume in all chambers in that head.

The prudent plan at this point is to remove slightly less metal than the amount indicated by the micrometer. By slight, we mean 0.002 to 0.004-inch. If too much metal is removed from the head and an overall lower chamber volume is produced, the head may be illegal when running in certain classes of competition. The head may now be taken to a competent automotive machinist. Tell him how much should be removed from the head surface. If at all possible, stand and watch—making certain that the head is jigged absolutely dead level on the mill table.

If all steps have been followed correctly up to this point, the chambers will now hold equal amounts of fluid; thus one less variable to high, consistent performance exists.

On the stem side of the cylinder head some of the valves are now "longer" than the others since they've been sunk deeper into the chamber. Changing the effective length of the valve stem alters valve timing and lift. This is a sure horsepower loser, and can negate all of your previous labors with the cylinder head.

Bolt a machined plate to the stem side of the head to serve as a resting place for a dial indicator. By moving an indicator along this plate and stopping at each valve stem, the relative height differential may be found. Measure them all—determine which is "shortest," mark valves as to location, and bring all of the rest of the valve stems down to that level. Do this on both heads. Most all valve grinders have an attachment for refacing the end of a valve stem—which makes the job of shortening a stem relatively easy.

The chamber-volume measurements used in this explanation are actual measurements from a 427 Chevy head. Naturally, for a variety of reasons, all heads will not measure the same—the figures used here should be considered only as an example.

When CCing piston-dome volume, locate piston 1.000 inch from deck if deck clearance is zero. Lightly grease rings and piston edge for a seal. Coat deck with grease before installing plate with filler hole to one side of cylinder bore.

FIGURING COMPRESSION RATIO

These photos, drawing, descriptions and formulas take the mystery out of computing compression ratios.

Don't measure cylinder volume—compute it from the bore and stroke and mark it V_1.

Cylinder volume =
  $0.7854 \times bore^2 \times stroke$

Deck-clearance volume V_2 is the volume between the piston top and the top of the block at TDC . . . if you are using flat-topped pistons. A micrometer, caliper or feeler gage can be used to measure the deck clearance and this measurement is figured as

Deck-clearance volume =
  $0.7854 \times bore^2 \times deck\ clearance$

Gasket volume V_3 can be estimated by assuming that the gasket opening is round and computing the volume:

Gasket volume =
  $0.7854 \times bore^2 \times gasket\ thickness$

But, the dome on most big-block pistons complicates figuring the c.r. And, for some classes, dome volume cannot exceed a specified amount or the engine will be ruled illegal. Therefore, you may need to know how to measure dome volume. The maximum number of cc's in the dome and minimum number in the chamber will give the most compression, of course. But, it is best to stay under the maximum compression by 1 to 2 cc's to allow carbon build-up without getting the combination into the area of illegality. If the pistons which you are using have too much dome volume, all can be reduced by milling off the dome. Be sure to get the valve notches correct before measuring dome clearance because such notching can remove a lot of cc's from the domes.

If a plastic or metal ring of known volume is sealed to the block over a cylinder, which has its piston at TDC, the ring can be filled from a burette. Subtracting fluid used from the ring volume gives exact dome volume *if the deck clearance is zero or above the deck.* If the deck clearance is below deck, the volume of the dome given by this method will be less than actual dome volume. The figure will be "off" by the volume between the flat surface of the piston and the deck at TDC. However, this method makes the compression ratio calculation faster and perhaps more accurate because you do not have to guess at deck-clearance volume V_2.

If a machined ring is not available, accurately locate the piston one inch down the bore. Use a plastic piece to seal the opening and measure the amount of fluid required to fill the cavity. The measurement is subtracted from the volume of a one-inch tall section of the bore. As with the ring method, this gives the exact piston-dome volume *if the deck clearance is zero,* but is off by the amount of the deck clearance if the deck clearance is below the deck. If your piston stands out of the bore or above the deck, the exact volume will be obtained if you move the piston farther down the bore by the amount that the piston protrudes from the block.

If your piston is below the deck, the exact volume of the dome can be obtained if you move the piston down the bore one inch *minus the deck clearance.*

Chamber volume V_4 is measured as described elsewhere in this chapter. If you are using domed pistons, the piston dome volume must be subtracted from the measured chamber volume to get V_4 for figuring compression ratio.

NOTE: Cubic inches from your cylinder volume are converted to cubic centimeters by the formula

Cubic centimeters =
  cubic inches $\times$ 16.4.

Just crank these figures into the following formula:

Compression ratio =
  $$\frac{V_1 + V_2 + V_3 + V_4}{V_2 + V_3 + V_4}$$

Remember that cylinder displacement or volume V_1 affects compression ratio. Anything you do to increase V_1 (bigger bore and/or longer stroke)—or to reduce V_2 or V_3 or V_4—automatically increases the compression ratio. If you plan to rework your cylinder heads now for use on an engine which will be bored and stroked later, use the displacement of the final engine configuration to determine the head volume that you'll need eventually so that you can do the head work just once.

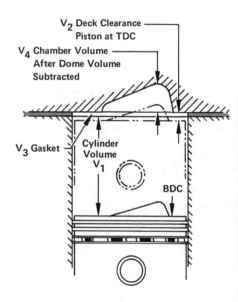

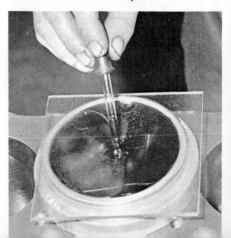

This shows one method of checking piston-dome volume. The piston is installed with rings at TDC. A machined ring of known volume is placed over the cylinder and all fluid measured as it is dribbled in. Subtracting fluid used from ring displacement gives the dome volume if the deck clearance is zero. Eye dropper is filled from burette and used to add critical last few drops.

Pistons
bigger is better

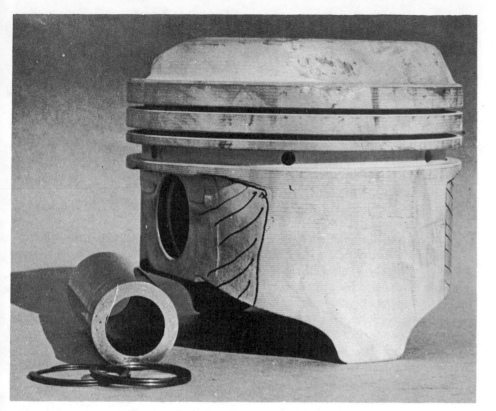

Chevy's forged oval-skirted full-floating-pinned piston. Black marks on piston show where oval surface connects flat side of pin-boss area into rubbing surface of skirt. Note that pin end is squared off. Clearance between pin end and Spiralox retainers should not exceed 0.005-inch when the pin is shoved hard against the opposite retainer. This reduces battering-ram effect of the pin and helps to ensure that the pin will not knock out the retainer ring.

STOCK PISTONS

Two basic types of pistons are supplied. Permanent-molded cast-aluminum ones are used in 396/402 engines of less than 375 HP and in 427/454 engines of 400 HP or less. Forged (impact-extruded) pistons are found in 375/425 HP 396's and 402's, 425/435 HP 427's and 425/450/460/465 HP 454's.

Several compression heights are available in the stock pistons. Compression height is the distance from the center of the pin to the top of the piston. Because the big-block pistons carry domes, or pop-ups, compression-height measurements typically include these bumps. But, if you measure the distance from the pin center to the piston deck, there are essentially only three compression heights. One accommodates the 396/402/427 engines, a shorter compression height (by approximately 0.120-inch) takes care of the 0.240-inch longer stroke of the 454 engines and one more handles the 366/ 427 truck engines with their four-ring pistons and taller deck height. There are three basic bore sizes in passenger-car engines: 4.094-inch for the 396, 4.125 for the 402 and 4.251 for the 427/454. The 366 truck engine has a 3.935-inch bore.

As you'll note throughout this book, pistons must mate with the cylinder head that you are using. There is a vast difference between the combustion-chamber volumes of the closed-chamber cylinder heads as used on all engines through 1970, except L-88 and ZL-1. All closed-chamber heads are mated to pistons with very small domes, including the first L-88's.

Open-chamber heads require larger-domed pistons to get back the compression that would otherwise be lost with the large chamber volume of about 117 cubic centimeters. And, as separately discussed in the cylinder-head section, there is no way to open up the stock closed-chamber heads to accommodate open-chamber pistons. Cast-iron open-chamber heads are available in both large- and small-port versions, so there's no longer any excuse for trying to convert a closed-chamber head to an open configuration.

The stock piston dome shape is head and shoulders above any trick dome shape that you can buy on a special big-block piston. Domes on the TRW replacements are essentially the same shape as those on the stockers. Chevrolet spent thousands of hours in development programs to mate the combustion chamber and dome shapes for the best performance. And, an enormous amount of effort was invested to make these combinations give good performance *with low emissions.* Even though the lower-performance engines introduced in 1970-71 don't appear to be all-out winners in the horsepower race, don't overlook the fact that the search for low emissions brought the open-chamber cast-iron head as a standard item for passenger cars ranging from the lowest-HP to the hi-perf models.

Steel struts cast into each side of cast-aluminum piston strengthen pin boss. A steel ring is cast into the piston to control expansion, thereby allowing close fitting and quiet running. Pin bosses are offset 0.060 inch toward thrust side to reduce piston slap. Cast pistons are perfectly adequate for most street-driven machines, including very hot ones. Valve cutout is adequate for highest lift stock hydraulic camshaft, can be deepened for special cams or Chevy's street-mechanical cam. Pin is held in rod by interference fit. Full-floating pin is not possible because there's not enough width to add retainer grooves to bosses. Lugs at each side are for balancing. Piston is slit or slotted below drilled oil-ring groove for additional expansion control. "S-5" atop piston is factory fitting code.

Last-forever, price-is-right forged piston for 396/402/427 closed-chamber heads. Uses pressed-in pins. Just the ticket for the budget builder. Available in 11:1 for 396 and 402 and 12.5:1 for 427.

Avoid trick dome shapes, including slots to "aid" flame propagation across the chamber, etc. Chances are good that you will never get the engine to produce as much HP with these specially shaped parts as you'll obtain with the stock-shape pistons. If you must buy special pistons, get them with domes shaped as closely to the stock product as you possibly can. This is one more of those areas in big-block building where GM's research investment can pay off for you ... so let it happen!

It's all too easy to convince yourself that you are building a racing engine when you are really after fast *transportation*. There's a big difference in what's drivable and what's not nearly that. So, keep your head on straight when you are buying pistons. There's no need to buy double-trick stuff if you're building a fast grocery-getter which will only see an occasional trip to the quarter-mile.

Unless you are building a strong racer, stick with the stock slotted-skirt autothermic cast-aluminum pistons. Get the forged ones if you are building an all-out screamer for drags, circle-track or top-speed events such as Bonneville. Cast pistons run quieter because they can be fitted closer—which gives you better oil control and longer ring life than you'll have with the forged pistons. Their compression ratio will undoubtedly be closer to that which will work with the poor-quality pump gasoline available today. So, here's another place that you may be able to save your hard-earned $$$ so that it will be available for the things that are really needed in the engine or for the chassis.

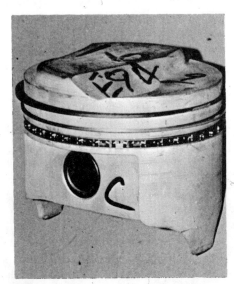

Stock Cast Pistons

These are of the autothermic design with a slot just below the oil-ring groove and a steel belt cast into the piston to control expansion. These pistons are typically fitted at 0.0007 to 0.0015-inch clearance, measured at 90° to the pin centerline. A wear limit of 0.0025-inch is specified, but wear limits become almost meaningless when you are after high performance. Chevrolet offers the cast-aluminum pistons in the standard bore sizes and in oversizes of 0.001, 0.020 and 0.030-inch.

Piston pins in the cast pistons are lubricated by holes in the top and bottom of the pin boss. The top hole is in a depression just above the pin boss so lubrication is by splash or throwoff.

Pins are retained in the cast pistons by making the pin an interference fit in the small end of the rod. There's no way to modify the cast pistons to use full-floating piston pins.

Stock Forged Pistons

Forged pistons (Chevy calls them impact-extruded) have solid skirts without slots. These are fitted at fairly tight clearances, typically 0.0036 to 0.0063 with a wear limit of 0.0055 to 0.0085, depending on the HP of the engine. Fits on factory-supplied forged pistons are typically a lot tighter than can be achieved with replacement pistons—with the exception of the TRW's. This is made possible by a barrel-skirt design which barrels the piston above the pin to a maximum of 0.0018-inch at deck height. This is done in combination with cam grinding and a tapered skirt which tapers outward 0.001-inch maximum at the bottom of the skirt. Forged-aluminum pistons are available in oversizes up to 0.060, as indicated in the Heavy Duty Parts List.

Piston-pins in Chevy's forged pistons are lubricated by oil scraped off of the cylinder wall by the oil ring. A hole at each side of the oil ring groove connects by another drilling into the top center of the wrist-pin boss on each side of the piston. Initially, all pistons were drilled from the bottom of the pin boss straight through the top of the pin boss and into the hole connected to the oil-ring groove. In 1970, the practice of drilling through the bottom of the pin boss, at least on the high-HP engines, was abandoned because this proved to be a weakening factor

Early L-88 pistons, for example, have the holes through the bottom of the pin boss, but examination of the later L-88's, ZL-1's and LS-6's shows that only the top of the pin boss bore is drilled at an angle so that the bottom of the pin boss can be left solid. "Little things" like this show that the continuing development program behind the big-block benefits every enthusiast who'll just pay attention to what those factory cats are up to.

Pins are retained in the forged piston in one of two ways: (1) the pin is pressed into the rod eye and held there by an interference fit in all of the cast-aluminum-pistoned engines and in some of the forged pistons: and, (2) the pin is full-floating in some of the forged pistons and is retained by Truarc retainers in early models and Spiralox retainers in later models.

TRW FORGED PISTONS

Replacement forged pistons are available from TRW in oversizes up to 4.375 inch. Their catalog lists pistons for both open- and closed-chamber heads for all of the passenger-car big-block versions. They even have an open-chamber piston for the 396 in standard bore size and in oversizes to fit the 402. Unfortunately, at least for those building engines for the street, the compression ratios on all of TRW's slugs are 11:1 or higher—except for one 10.5:1 set for the 396 closed-chamber heads. Domes are designed with the required thickness, but they can be trimmed to reduce compression. A minimum dome or head thickness of 0.200-inch can usually be used with complete safety. If the engine is being built strictly for street use and you are having problems coping with "no-octane" gas, a dome no thinner than 0.180-inch can be run—*on the street only!*

Some builders feel that TRW's shallow grooves in the pin bosses will not oil the pin correctly. You can look at the stock piston to see how the pin is oiled, then do something similar to your TRW's. Above all, don't make the common mistake of drilling holes through the bottom of the pin bosses with the idea that this will provide the increased oil that the pin needs. It not only won't provide the lubrication that's needed—it can weaken the piston at this highly stressed point.

Another thing the long-distance racers sometimes change on TRW pistons is

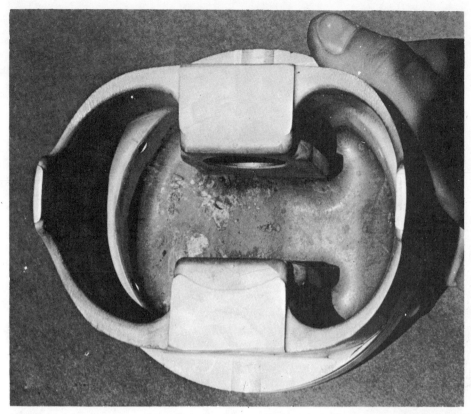

Underside of stock Chevy forged piston shows oval-skirted configuration which was used for open-chambered L-88's and ZL-1's and as a HD Parts List item for building an LS-7. This design was dropped in favor of the straight sided design. Undrilled pin-boss bottoms indicate that this is at least a 1970 or later piston. All of these pistons with floating pins are equipped with 1/16-inch-thick compression rings.

the 0.050-inch-wide retainer grooves. These can be machined to accept the 0.072-inch-thick retainers as used by Chevrolet.

If you use Truarc retainers, install the sharper edge of the retainer toward the cylinder wall so that the Truarc will "bite" into the piston groove for better holding power.

Pin buttons can be used, of course. These are preferred by many drag racers.

Two drills have been inserted in the pin-boss area of this forged Chevy piston. Note how oil from the back side of the oil ring connects to angled hole drilled into pin boss. Similar holes can be added to the TRW pistons for improved pin oiling.

Pen points to area where some Chevy pistons have an oval-skirt design. Black lines help to provide a comparison between the straight and oval skirt. Straight-skirt TRW is shown here. 1971 ZL-1, L-88 and LS-7 12:1 pistons and 1st design L-88's also have the straight design. Oval-skirt pistons do not always have a symmetrical skirt and some builders remove metal from the wider contact area to make the rubbing surfaces symmetrical.

HOTROD PISTONS

The only time that you should consider using hotrod-type pistons made by the custom piston manufacturers is when stock Chevrolet or TRW replacement parts are not available in the bore size or stroke that you need. Hotrod-type pistons cost more money than the Chevy/TRW parts but we doubt that you'll find any more performance. Of course, if you are building some special-stroke engine—other than the Chevy or Reynolds Can Am-type block—you will probably have to use pistons made by Arias, Forged True, Jahns, J.E., Venolia, etc. However, you may want to give serious thought towards using the stock stroke and oversize bores not exceeding 4.311-inch (0.060 over 427/454) so that you can use the fine high-quality stock or TRW pistons. A few extra inches will not usually provide enough extra HP to be worth the extra time and $$$ required for the hotrod pistons. Obviously, if you are building a giant "moose motor," special pistons will be required.

If you buy custom pistons, allow plenty of time for delivery because these typically take from several weeks to several months for the manufacturing process, especially if the size and compression height that you need is not scheduled into a production run of some type. Don't plan to build an engine on a double-quick overnight basis with special pistons. That's frustration-ville! Get the pistons in your hands before trying to establish a time schedule in which to complete your engine.

DISPLACEMENT TABLE

Bore & Stroke = Inches   Displacement = Cubic Inches

Stroke \ Bore	3.935^A	4.094^B	4.125^C	4.155^D	4.187^E	4.251^F	4.281^G	4.311^H	4.341^J	4.376^K
3.47^1	338	365	371	377	383	394	400	405	411	417
3.76^2	366	396	402	408	415	427	433	439	445	452
4.00^3	389	421	428	435	442	454	461	467	474	482
4.25^4	413	447	454	462	469	482	490	496	503	511

1. Can Am crank 3993803 (limited availability)
2. Stock 366/396/402/427 crank
3. Stock 454 crank (must be used with externally balanced harmonic damper and flywheel)
4. 454 + 1/4-in. stroke, 427 + 1/2-in. stroke

A. 366
B. 396
C. 396 + 0.030 or 402
D. 396 + 0.060 or 402 + 0.030
E. 396 + 0.093
F. 427/454
G. 427/454 + 0.030
H. 427/454 + 0.060
J. 427/454 + 0.090
K. 427/454 + 0.125

NOTE: To calculate displacement, use the formula: 0.7854 x Bore x Bore x Stroke x 8 = CID

FORGED — OR CAST?

Some stock and replacement pistons can be used for racing, but not all racing pistons are suitable replacements for the stock items. Suitability of the various types depends on the intended use for the engine. Many replacements and racing pistons are available for the big block, including sand-cast, permanent-molded, die-cast and forged types. Of these, the forged piston is the ultimate for several reasons. Forged pistons are made by forcing aluminum slugs into a piston form under extremely high pressures to create a very dense grain structure in the finished part. Forged pistons are claimed to have up to 70% more strength and better heat-dissipation characteristics than conventional sand-cast pistons. Their strength at temperatures over 600°F is far superior to any other type of piston.

TRW, which makes more forged pistons than any other company in the world, has this to say about forged piston qualities and capabilities. "Forging starts with a billet of alloyed aluminum which is preheated to a working temperature and then formed. A multistage forming process allows control of the internal grain flow in the head, skirt, pin boss and ring land areas. The resultant piston blank has exceptionally high density with virtually no porosity."

Sustained full-throttle, full-load operation substantially reduces the hardness and strength of the cast piston. The density and section shapes which are possible in the forged piston greatly reduce the temperatures in these critical areas, giving the forged piston a considerably higher operating strength range which makes it more resistant to head or ringland distortion.

Several makes of pistons are available for the big block, and most of these come complete with pins, rings and cylinders—except the special pistons which are available from custom manufacturers.

Although forged pistons have a number of advantages over die cast and permanent-molded pistons, the initial investment required for the forging dies, and the use of more expensive production machinery, has kept most manufacturers from ever considering this method of manufacturing pistons specifically for the big block.

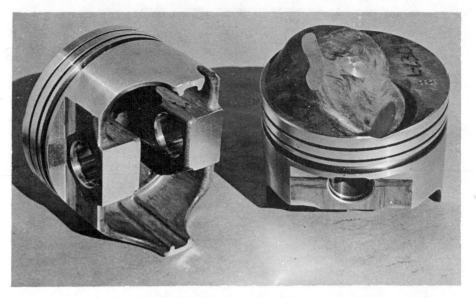

Inside and outside views of TRW forged pistons for 427 and 454. Note ribs inside of skirt and flat sections connecting pin-boss areas to skirts. Shadow indicates the deep valve reliefs which TRW provides in these pistons. These are for open-chamber heads. TRW also has a standard-bore 396 piston designed for use with the open-type heads.

Almost any of the pistons — whether forged or cast — will work fine for road or street use. All that you have to do is to make sure that the pistons are fitted with the correct clearances. Watch the mixture strength so that you don't run the engine too lean and don't try to crank in more than the recommended 38° to 42° BTDC maximum total advance.

Ribbed skirt and squared off ribs to pin boss are features of TRW forged pistons.

MAXIMUM BORE SIZES

It's no secret that the 396 and 402 blocks can be safely bored to 427/454 size of 4.251 inches. And, Chevrolet sells 0.060-inch oversize pistons for this bore—so you can probably figure on this working out safely in the 427/454 block. Bore sizes larger than this in any of these blocks can be accomplished—*at your own risk*. We do know that some 4-bolt 396/402 blocks have been bored to as large as 4.311 inches, but not always successfully. We have heard of 427/454 blocks being bored 0.125 to 4.376 inches. If you want a 4.44-inch bore, try to find a Can Am block. Chevy has two pistons for that bore size to cover three different strokes.

Hotrodders learned long ago that cubic inches are the only absolutely trouble-free hop-up trick. They know that bigger bores are the logical way to go. As Tom Medley (publisher of Rod & Custom Magazine) said way back in 1947, *"When in doubt, bore it out!"* Fortunately, the big-block Chevy has unusually thick cylinder walls. These are nominally 0.260-inch thick (if there's no core shift) in high-performance 396 blocks.

The 366 block is a thin-wall casting and anything over 0.125 inch overbore will be getting risky.

Aluminum ZL-1 blocks can be overbored 0.030 inch.

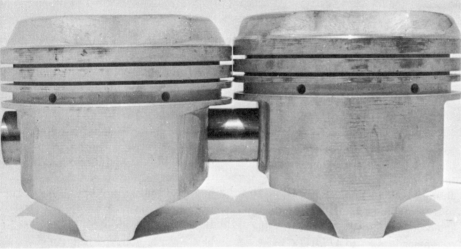

Top and side views of an LS-7 piston for 4-inch stroke and open-chamber head compared with an L-88 first design piston for 3.76-inch stroke and closed-chamber head (right). Note that the LS-7 454 piston has the pin nearer to the flat surface of the piston and the skirt is shorter. Both pistons are forged with full-floating pins. S4 and S-6 on top of pistons are factory codes indicating the diameter within a tolerance range to allow quick fitting of the engines when they are built. These are stock pistons from the HD Parts List.

Undersides of two Chevrolet forged pistons. LS-7 454 at left has no oiling holes through bottom of pin bosses, is equipped with 0.072-inch-thick Spiralox retainers. 1st-design L-88 piston has 0.050 Spiralox, drilled boss bottoms and straight-sided skirt-to-boss construction.

COMPRESSION RATIO

Dome size relates to compression ratio, of course, as does combustion-chamber and cylinder volume. When you bore the engine for larger pistons, the dome must be made smaller—or the combustion chamber increased in volume to keep the same compression ratio provided by the stock combination.

Keep in mind the fact that gasoline quality has been getting worse in most areas of the country. On the West Coast, the last good super-premium was taken off of the market in 1970. We understand that the problem is similar throughout the U.S. Of course, you can buy aviation gasoline for racing purposes, but your daily transportation needs the ability to run on what's available at any corner pump.

Use a compression ratio of 9.5 to 10:1. If you insist on using 12:1 because Chevy and TRW offer such pistons, you'll "go bananas" from your engine's continual rattling on low-octane gasoline. It's not smart to build an engine with a lot of compression for street use anymore—unless you are using LPG fuel.

PISTON PREPPING

Before you do anything else to your new pistons, measure them for bore size and compression height. There's no sense investing a lot of preparation time on pistons that are wrong to start with. It's happened!

It is a very good idea to get your new pistons Zyglo-ed before you start deburring the domes. New pistons, even though they are forged, are not always perfect. Sometimes the metal does not flow into all of the places that it should have and any discontinuity could be the beginning point for an eventual crack.

When you get your pistons, go over the heads with sandpaper and files to get rid of all sharp edges.

PISTON FITTING

When you buy pistons for your engine, get them in your hands before you bore or hone the block to size. Measure each piston with a micrometer and check each bore with a snap gage and micrometer. Mate the pistons to the cylinders so that there is the least amount of honing to be done in each of the cylinders. You can also check the piston compression heights to see whether a certain combination of compression height, connecting-rod length and crankshaft rod throw will assist you in getting the best possible tolerance stack up for the right deck height at each cylinder. When boring to fit a class limit, stay three to five thousandths undersize to allow for wear or rehoning. Otherwise, it will be new-block time before you'd like to be there. Or, your engine could be declared to be oversize in a tech inspection teardown.

Pistons can be fit as closely as 0.0045-inch in the big block with little danger of scuffing—but the engine will not develop full HP until you've run it for about 30 to 50 hours under load. However, the long-term durability of the engine is improved because full HP is not being developed and the pistons do not rock as much in the bores. When pistons are fitted according to the factory hi-perf clearance recommendations (0.0065 to 0.0075-inch according to the table in this book) long-distance racers often find one or more pistons cracked at the end of a race. For this reason, any big-block used in long-distance racing must always be torn down and fully inspected, including Zygloing the pistons, after each race event. The same is not true for circle-track and drag racers, because they do not spend extended periods of time at peak RPM's under full load. However, Zyglo-ing is recommended whenever the engine is torn down for rings, etc.

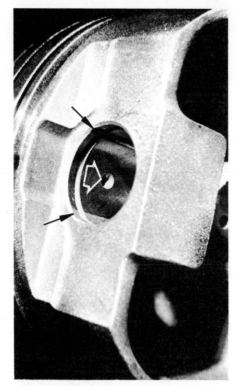

Drilled TRW piston shows new oil holes for the pin bosses (outline arrow) and grooves which TRW supplies for pin lubrication (black arrows). Interior has been glass-beaded and the exterior has been beaded around the wrist pin and on the head—but not on the thrust surfaces. Photo courtesy King Engine Service.

You may as well figure on getting all new pistons after every long-distance road race.

If your fire breather will live on the trailer and you're all set to change rings after about 30 hard passes through the quarter mile then you can get away with 0.0075-inch. Running 0.0075-inch piston-to-bore clearance on the street in any large-block Chevy is an engraved invitation for a noisy, oil-consuming, plug-fouling, short-lived engine. You'd hate it!

It's not uncommon to find four to six pistons cracked in the pin-web area when pistons are Zyglo checked after a long-distance event. Pistons and con rods must be considered as expendable replacements when racing the big block. Don't be cheap or hesitant when it is time to replace parts, especially if cracks or flaws are detected prior to assembly. And, it is much cheaper to find them then!

Sharp edges on notches and dome of this piston must be rounded off carefully before piston is run. "Trick" dome shapes which are vastly different from stock should be approached with caution. If you have the money to buy them, dyno them in comparison to the stock pistons — and then be able to throw the trick stuff away away without flinching when it doesn't make as much HP as the stock stuff — then you are ready to become a true racer. Dial indicator is being used here to locate TDC. Magnetic base is holding onto one of the iron liners in the block.

PISTON CLEARANCES

Type	Road Use	Street/Strip	Racing
Chevrolet cast-aluminum	0.0015-0.0025	0.0025-0.003	0.003-0.004
Chevrolet forged	0.004-0.005	0.005-0.006	0.0065-0.0075
Hotrod cast or forged	Follow maker's recommendations		

NOTE: Measurements of piston should be made perpendicular to the wrist pin, even with wrist pin centerline.

PIN CONFIGURATIONS

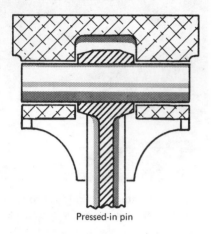
Pressed-in pin

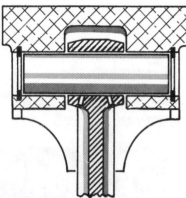
Full-floating pin with retainers

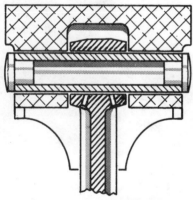
Full-floating pin with pin buttons

These three drawings show three ways that pins are installed in big-block engines. The top two are stock methods and the bottom one is favored by some racing-engine builders. Note that the two full-floating methods have pin clearance in the rod eye and two holes in the rod provide pin lubrication.

NOTE: — Stock L-88/ZL-1 rods usually give 0.0003–0.0005-in. pin clearance and must be honed to the recommended 0.0005–0.0007-in. When reading about pin retainers, remember that excessive rod side clearance at the journal seems to aggravate pin-retainer durability.

PRESSED-IN vs. FULL-FLOATING PINS

Two types of piston pin installation are used in Chevrolet big-block engines. Most common is the pressed-in pin. In this situation, the pin oscillates in the piston pin bores, but is pressed immovably into the small end of the rod with an interference fit. The vast majority of all Chevrolets on the road today are so equipped. High-performance big-block engines such as L-88 and ZL-1 have full-floating pins. The pin is a floating fit in the piston-pin bores and in the small end of the rod. Spiralox spiral-wound steel retainers, either 0.050- or 0.072-inch thick, hold the pin in the piston. All other big blocks, including those with forged pistons, have pressed-in pins. These pistons do not have retainer grooves and there's no room to install them.

Full-floating pins are recommended for any engine that is to develop a lot of HP on a continuous basis. The reasons include longer life for the pin bores in the piston, for the pin itself, and for the rod— if the rod has adequate pin clearance and both rod and piston have suitable pin-oiling holes. The advantage of easy assembly and disassembly without danger of damaging the parts is especially useful for a racing engine which must be disassembled and reassembled often because it is being raced.

In a high-loading situation, a pressed-in pin tends to scuff in the pin bore because the pin is marginally lubricated and the same area always gets the load. Once the pin starts to scuff, the pin bore overheats and rapidly wears to an out-of-round configuration. Then the rod can cock and break in the middle. The same type of failure occurs if a full-floating pin is not lubricated in the piston or has inadequate clearance or lubrication in the rod. In this situation the pin galls or scuffs in the rod eye, seizes so that it no longer turns, then scuffs in the piston with a failure sequence identical to that of a pressed-in pin which scuffs in the piston pin bore/s. Such failures are nearly always blamed on the piston or on the connecting rod—instead of on the pin which was the real culprit.

Full floating pins should have 0.0004 to 0.0008-inch clearance in the piston; 0.0005 to 0.0007-inch in the rod. If the fit is made too tight, the broken-rod syndrome could occur in *your* engine.

As the rod moves back and forth on the crankshaft journal there is a tendency for the pin to move back and forth with the rod. In a full-floating configuration this makes the pin act as a battering ram which tries to push the pin retainers out of their grooves. Because this happened on occasion with the 0.050-inch-thick Spiralox, Chevrolet increased the retainer thickness to 0.072" to hold the pins in place. Keeping the pin-to-retainer clearance to a minimum (0.000–0.005-in.) preferably the zero clearance — and setting the rod side clearance at no more than Chevy's specs — reduces or eliminates this battering-ram effect. If you machine or remachine grooves in a piston to use the thicker Spiralox, this must be done with great care to end up with the desired end clearance. If you are changing to a full-floating setup, pin bores should be opened up to provide the recommended pin clearance. Pin bores can be drilled to provide lubrication for the pin and the rods must be modified for use with full-floating pins.

Retainers should be used *once!* After you have run the engine, use new retainers if you take them out of the piston for any reason. Truarc retainers are supplied in many replacement pistons. These are apparently adequate for most applications, but builders seeking maximum reliability often use two Truarcs at each end of the pin in wider grooves — or use the thicker Spiralox in wider grooves.

There's sharp disagreement among engine builders as to the relative merits of pressed-in versus floating pins. Engines are easiest to assemble when floating pins are used. But, some engineers insist that the engines will live longer and provide less trouble if the pressed-in pins are used. They point out that the pressed-in pin provides a "T-shaped" structure for the top of the rods so that the pistons are prevented from cocking on the rod axis, as occurs with the floating pins. This gives the pin only two bores to "wiggle" in, instead of three. There's no denying that correctly installed pressed-in pins are best for a street engine. The engine will stay quieter longer with the pressed-in pins than with the floating pins. However, pressed-in pins can be installed incorrectly and when this happens, disaster is certain to occur. When fitting pressed-in pins, it is sometimes necessary to hone the rod "eyes" or small ends so that the interference fit for the pins will be reduced to an acceptable level. Too much interference galls or tears the rod eye or pin surface as the parts are pressed together, causing very rapid wear of the pin-bore in the piston. The interference fit should be in the range of 0.0008 to 0.0012. The tighter fit is preferable.

It is extremely important to ensure that the machinist uses the correct fixtures and measures the pistons, rods, and pins—and then corrects any misfits before proceeding with the installation. More pistons are ruined by faulty assembly to the pins and rods than by any other cause. After the correct fits have been established, use the right fixture to center the pin and rod in the piston. The fixture also guides the pin through the rod eye and into the other pin bore. The pin and rod eye must be lubricated prior to pressing the pin into the rod/piston assembly. And, the piston must be supported squarely and firmly. Some machinists heat the rod eye prior to installing the rod onto the pin and piston. Pin clearance in the piston should be 0.0003 to 0.0005-inch to ensure good oiling. Some replacement pistons from Chevy may have as much as 0.001-inch clearance (upper tolerance limit).

TAPER-WALL PINS

You'll note that some of our photos show taper-wall pins. These weigh slightly more than stock pins, but they concentrate the major cross-sectional area of the pin where the load is greatest—in the eye of the connecting rod. The ones shown are made from 9310 bar stock, are specially heat-treated and have squared ends to allow close fitting to the retainers. Stock ZL-1, L-88 and LS-7 piston pins also have squared ends. Some of the piston manufacturers also offer the taper-wall pins. The piston pins are usually available in a standard length for use with Spiralox or Truarc retainers and longer pins are offered for those builders who prefer to use end buttons.

Taper-wall pins were designed specifically for big-block racing. Do not get any that are chrome-plated because any irregularities in the chrome will cause the pin to break.

8.5:1 piston from 1971 454. Use of low-compression pistons is going to spread as the gasoline quality continues to sink. Note slotted oil-ring groove, valve notches and balancing pads. "Groovy" skirt finish is just the way Chevy makes 'em. This is a cast-aluminum piston designed for use with pins which are pressed into the rods.

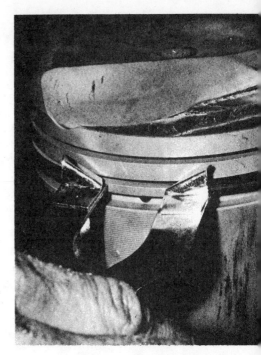

Install rings with a ring-expander tool. If you twist the rings into the grooves, the rings may take a permanent "set" which will cause early failure. Photo courtesy Popular Hotrodding.

Take it easy when you install those rings. Ring compressor must be flush with the block deck at all points before driving the piston down with a piece of wood.

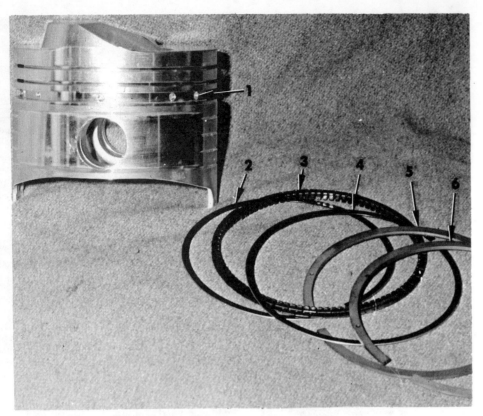

Holes 1 drilled in the oil ring groove allow oil to return to the sump. 2 is the lower rail for a three-piece oil ring. 3 is an oil-ring expander. 4 is the upper rail. 5 is the second ring. 6 is the first or compression ring. The dot on the top of the ring should always go up on the piston.

Bill Jenkins takes a close look at a piston ring with a lighted magnifying glass. How many rings have you ever examined by this method? Might learn something. Same illuminated magnifier is used to "read" spark-plug electrodes and porcelains.

PISTON RINGS

Piston rings are always a subject of interest among auto enthusiasts because they are almost always replaced during engine rebuilds and must, of course, be replaced if the bore size is changed. Few mechanics will hesitate to debate the merits of various brands of piston rings, but many do not understand the basic functioning of these vitally essential parts.

The sealing of combustion and compression pressures occurs as these same pressures enter the ring grooves, forcing the rings down against the bottom of the grooves and outward against the cylinders. Although outward stresses are built into the rings, these merely *assist* the gasses in sealing. Ring pressure against the wall does not accomplish much sealing. Tests made by Ricardo showed that rings operating *without* gas pressure behind them failed in a few minutes of operation.

Racing-engine builders frequently talk of "ring flutter." At high RPM the top ring may have sufficient inertia as the piston decelerates near TDC, so that it is held against the top of the ring groove. Gas behind the ring escapes to the crankcase and the top ring collapses *inward* under cylinder pressure, instead of being pushed outward as in normal operation. This results in vastly increased blow-by which pushes oil out of the crankcase, and drastically reduces power output. Ring collapse or "flutter" occurs only on the power stroke and always causes ring breakage if allowed to continue.

Designers of modern high-performance, high-RPM engines avoid ring flutter by reducing compression-ring width, thereby decreasing their weight and consequent inertia. Ricardo's famous book says, "Taking all of the arguments into account it would seem that the width of the ring should depend on the normal speed of the engine; the higher the speed, the narrower the ring." As proof of this comment we see high-speed two-cycle engines such as racing outboard and motorcycle engines using 1/32-inch-wide rings to prevent flutter at speeds which may exceed 10,000 RPM.

Any compression ring thicker than 1/16-inch or one-piece oil rings thicker than 1/8-inch are probably too thick to provide any oil control whatsoever in a high-RPM engine. Rail-type oil-ring setups which are combinations of very thin rings with wavy spacers can be wider, of course. If you have pistons with wider ring grooves, you might investigate Perfect Circle's spacers which allow using thin rings in wide grooves. Don't waste your time trying to build huge breather boxes to contain the oil from a bad blow-by condition. You will be way ahead horsepower-wise by eliminating the blow-by problem through the use of the correct rings.

A trend began to grow very fast in 1963-64: racing-engine builders turned to a 20-year-old idea to solve their ring-flutter problems on big-bore high-horsepower engines. They began installing an L-shaped ring invented by Dr. Paul de K. Dykes of Cambridge University. These rings are used almost exclusively in high-performance engines. Even so, because the rings use low tangential pressure against the cylinder wall, cylinder wear may be reduced during those strokes when the rings are not pressure-loaded. Dykes-type rings require a special stepped groove or a spacer can be inserted in a standard, non-stepped groove to allow using them in a piston which was not originally machined for Dykes rings.

Although there are many small-block Chevys and other engines equipped with the Dykes-type rings, we did not find that there had been much experimentation with these for the big-block engines. This is perhaps due to the fact that the moly-filled rings began to be used as stock equipment about the time that the big-block came onto the scene. This drastically reduced the price for the moly rings and made them easier to obtain. And, the 1/16-inch thick compression rings typically used in the big block are apparently thin enough to provide flutter-free sealing at the RPM limits typically used for these engines. Part of the reason for the lack of use and limited experimentation with Dykes rings for the big-block lies in the cost. Such rings are more expensive to make in large bore sizes and you'd have to pay several times the price of a set of normal chrome-plated or moly-filled rings to get a set with Dykes top-compression rings.

Ring Types

Moly rings - These are recommended because they seal instantly without any break-in time if the cylinder is round and smooth, even if the bore is used. Chevrolet claims that the moly rings give less friction than other types of rings, also. The instant seating is provided by a combination of features: (1) the cast-iron channel on each side of the moly inlay, and (2) the smooth surface grind on the ring face. When moly rings are made, they are lapped in a round cylinder until they are *light-tight*. Such processes are not used for ordinary cast-iron or chrome-plated rings.

Some makers enhance the sealing characteristics with a tapered face or a controlled torsional action to make sure that the ring will provide the initial hairline contact of cast iron against the cylinder wall. Moly rings are not pure molybdenum, but start out as cast-iron rings with a channel. This channel is filled with molybdenum-spray which molecularly bonds itself to the cast iron. About ten thousandths of cast iron is left at both the top and bottom of the ring.

With a melting point about $1000°F$ higher than chrome, moly is very resistant to scuffing. This scuff resistance is augmented by microscopic cavities in the moly coating. These fill with oil to give excellent pressure sealing and scuff resistance.

The moly rings are used only in the compression positions on the piston. The usual chrome-rail with wavy expander oil rings are typically used in moly ring sets. Chevrolet's moly ring sets have moly top compression rings. TRW, Perfect-Circle and Ramco sell "double-moly" sets with moly 1st and 2nd compression rings.

Regardless of who made the rings, examine the surfaces carefully for chipped or damaged surfaces. Reject any that are not right because there is no sense installing bad rings in your engine to start with.

Chrome-plated rings - Chrome-plated rings were the "hot tip" before the advent of moly-filled rings. And, in some instances, the extra abrasive-wear resistance of the chrome rings makes them the only ring to use. Chrome-plated rings gained widespread use because they are four times harder than uncoated cast-iron rings and therefore they cut abrasive-caused wear by as much as five times. Chrome-plated rings are therefore the best buy for any form of racing where the engine is not equipped with an air-cleaner system. Chrome rings will live approximately twice as long as moly rings in such applications.

You should only consider chrome rings if there is no way to run an air cleaner and if your round cylinders have the correct surface finish. Chrome rings cannot conform to an odd-shaped cylinder and may never seal in a worn cylinder. Hence, they are seldom used for re-ringing unless the cylinder has been honed round.

Chrome rings must be carefully inspected for chipped or damaged surfaces prior to their use or reuse. Such defects can cause serious cylinder-wall damage if the defective rings are used in an engine.

CAUTION ON GLASS BEADING — If you glass-bead pistons, mask off the ring grooves before you start. Then be sure to keep the high-velocity stream of beads away from the ring-groove area. Beading can destroy the side seal on rings.

Enlarged view of stock Chevrolet moly-ring set shows top compression ring with cast-iron channel edges at each side of moly-sprayed center filling. 2nd compression ring is plain cast iron. Oil ring has chromed rails with wavy separator.

Checking groove depth or back clearance. When ring is bottomed, its depth should closely equal that of the groove in which it will operate.

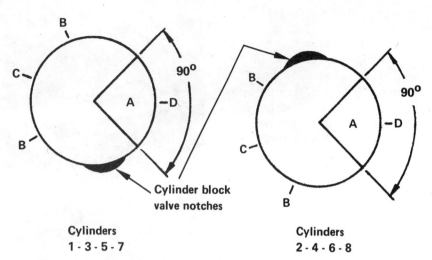

Here's how to install big-block piston rings on the pistons. A is the arc within which the oil-ring-spacer gaps should fall. Put the tang in the hole within this arc. B is the oil-ring-rail gap. C is 2nd compression-ring gap. D is top-compression-ring gap. Be very careful not to get any gap into a cylinder valve notch or you could tear up the cylinder wall with a broken ring. You would not feel the ring break in most instances, so the damage would occur and you would not know it until too late. This gap arrangement provides best oil control. This is the type of information that you will find in the Chevrolet Overhaul Manuals. You should have one to work on the engine with the assurance that you are doing things correctly.

Oil Rings

Oil control is a separate subject. Dr. Dykes has also been responsible for research which proved that oil passes the rings in both directions and in large quantities. Although oil flows past the oil-control rings on the down-stroke, it is returned to the sump on the upstroke. There are at least two types of oil-control rings to consider: (1) the type using two chrome-plated rails above and below a wavy ring separator, which probably should not be used for re-ring jobs, and (2) the dove-tail design slotted cast-iron ring with an expander.

Fitting Rings

In addition to the usual dimensioning of the rings to fit bore diameter and ring-groove width, three other measurements are quite important: ring end gap, groove clearance and "back" clearance.

End gap is measured by pushing the ring squarely into the cylinder about an inch with a piston. A feeler gage can be used to measure the end gap, which should be not less than 0.0045-inch per inch of bore. Example: For a 4.25-inch bore, 4.25 x 0.0045 = 0.0192—or 19-thousandths of an inch. It is better to err on the high side and get too much gap than too little. A gap of 0.019 to 0.022 is suitable for this bore size—and a gap of as much as 1/32-inch would not affect performance. End gaps which measure too small when the ring is installed in the cylinder can be opened up by holding the ring in a vise, protecting it with a cloth or cardboard, and carefully filing or stoning off the ends to get the desired gap. If your pistons arrive with the rings installed, don't be lazy —take them out and measure the ring gaps and the groove clearance as discussed herein and in the service manual.

Ring-groove clearance is also measured easily with feeler-gages. Stock-piston-groove clearance for compression rings should be from 0.0012 to 0.0032, measured between the ring and land. Oil-ring clearance, measured between ring and land, should be 0.002 and 0.0035. Replacement and racing pistons often specify 0.002 to 0.004 clearance for compression rings. Special groove clearances are needed for Dykes-type rings. Ring-groove clearances on the low side improve oil control and high-speed performance, but increase the risk of ring sticking in a short period of time, especially where engine is used only on the street. Too much groove (side) clearance increases oil consumption, causes groove pounding, and finally ring breakage. Additional side clearance can be obtained when needed by sanding the side of the ring on a flat surface. Solvent-wetted Wet or Dry paper supported on a thick piece of glass works nicely.

"Back" clearance is the clearance between the groove "bottom" and the backside of the ring when the ring is at cylinder-bore diameter. The back clearance is o.k. if the ring width and groove depth are nearly identical—or perhaps the groove could be 0.010 to 0.030-inch deeper than the ring was wide. A common problem with oversize pistons was that the groove depths were left at the size which was correct for the stock bore, thereby providing an enormous amount of "back" clearance on oversize pistons. Ring-groove spacers have been made for years out of ring shim stock which was stacked into the groove after being formed into ring shapes. Enough ring shim stock was added to the top compression ring groove, and sometimes to the second compression ring groove so as to take up the unwanted clearance. However, this shimming procedure was time-consuming and not always satisfactory, especially if the shims stacked up unevenly to cause a bulge.

Bob Joehnck says that back clearance is no longer the problem that it once was because most oversize pistons are supplied with the correct ring groove depths and the oversize rings are made to provide minimal back clearance. If you are preparing a transportation machine, a quick check to see that the ring grooves are not excessively deep should be all that's required.

But, if you are building a drag-only engine, you may want to optimize the back clearance to get the highest possible compression pressures and perhaps longer ring life.

The current best way of shimming ring compression-ring grooves to reduce back clearance is to use a coiled spring which is placed behind the ring. The wire thickness which is used allows grinding the coil to a lesser thickness to make the back clearance exactly what the builder wants. One builder who has used such spacers is pro-stock racer Ron Hutter of Chardon, Ohio. He recommends using a back clearance of 0.015-inch per side, or 0.030-inch smaller groove OD than the ID of the ring when it is expanded against the cylinder wall. This leaves the ring standing out of the piston somewhat when it is bottomed in the groove. By using this back clearance, which is much tighter than many builders use, Hutter has claimed as many as 75 or 80 runs from a single set of double-moly rings.

After checking and adjusting ring end gap, measuring ring groove clearance and setting ring "back" clearance, direct your attention to the ring grooves themselves.

Ring Grooves

Ring-groove condition is vital to the functioning of the rings. According to Richard MacCoon of Grant Industries, "Rings are just like valves and must have a good seat on which to seal. Groove finish is every bit as important as that of the cylinder or the ring itself. Piston makers machine piston-ring grooves with extreme care to ensure perfectly smooth surfaces. Yet you will often see 'mechanics' use a broken ring to clean out carbon from these finely machined grooves. They gouge and burr the groove and still expect the rings to work. Special tools for reconditioning grooves should always be used when fitting new rings to used pistons."

Ring Installation

When you install the rings on the pistons, follow the instructions as to where you should put which rings. USE A PISTON RING INSTALLER to get the rings on the pistons without damaging the ring grooves or lands between the grooves . . . or the rings. If you don't own a piston-ring-installing tool, buy one! When the rings are on the piston, oil them and position the gaps in the relative positions indicated on the installation diagram. Do not allow any ring gap to enter the valve-clearance notch in the cylinder wall or you could end up with a totally demolished cylinder wall caused by a broken ring. It is possible to break a ring in the notch as you install the piston, so don't let this happen to your engine.

HONING

Many racing-engine builders use the CK-10 Sunnen Power Hone because the hone leaves the cylinder with the desired cross-hatch finish and the correct degree of smoothness to ensure that moly rings will seal correctly.

Chevy's own fine pistons are used to advantage when displacement is limited to stock size by class regulations, as in stock drag classes. And, these pistons should be at the top of your list if the available oversizes meet your displacement requirements If you plan to use the engine for high speeds and drags, more clearance is needed. See the Piston Clearance Table for details.

Add clearance to the stock bore by honing the cylinders. You do not have to remove the pistons from the rods when you take the assemblies out carefully (after covering the rod bolts with plastic tubing) so that the cylinder walls are not damaged by the rods. But, like the old Indian scout—look for signs. Check the wear pattern on each piston to make sure that it is parallel with the piston-skirt centerline. If not, this is your "sign" that the piston is cocked and not travelling straight in the cylinder. Any piston showing uneven wear must be aligned with its connecting rod. Twisted or cocked piston/rod assemblies steal horsepower through unnecessary friction and create unwanted heat and wear. You'll also want to look at the cylinder bores for similar signs.

When honing the cylinder block it is essential to simulate cylinder distortion caused by installing the cylinder head. There's HP to be gained by this attention to detail—so don't neglect it. It is worth the extra time that is required. Some racing mechanics bolt a plate to the top of the block—a plate with holes large enough to allow a hone to enter the bores—and which has the same bolt-hole pattern as a big-block cylinder head. We have seen some of these that are 2" thick. Chevrolet Engineering suggests that 7/16-inch NC capscrews with washers be installed in all of the head-bolt holes and torqued to 60 lbs. ft. The end result of this effort is that the bolts and/or plate simulates the stresses applied to the block by the cylinder head. The hone then makes a rounder hole . . . one which will be round when the engine is bolted together.

PISTON MACHINING
Skirt knurling and grooving

Numerous articles have appeared which illustrated engines being assembled with new pistons which were knurled to decrease clearance. Knurling, as often practiced in cheap engine rebuilding, deforms the piston skirt to decrease piston side clearance and thereby improve oil control and reduce noise from piston slap. Because of the deformation which occurs, new pistons should not be knurled on the skirts.

Some engine builders turn shallow grooves on the skirt of the piston to guarantee that the piston skirt will be lubricated by oil in these "reservoirs." Both the TRW and stock Chevy pistons have a wavy finish on the skirts for this reason.

Piston balancing

Any replacement piston will probably weigh more than the stock item —especially if you are increasing the bore size. This is not a problem so long as all piston and pin assemblies weigh the same and the crankshaft, connecting rods, and flywheel and clutch are carefully balanced by a capable and experienced balancing expert. Avoidance of excessive piston weight improves engine life by reducing loads on the piston itself, pin, connecting rod, bearings, and the connecting-rod bolts. Material should not be removed from the slipper or load-bearing portion of the skirts or the pistons will be seriously weakened. Balancing pads are provided to permit equalizing the weight of all pistons in a set.

Replacing less than a full set of pistons

Replacing a piston in an engine is not a quick and easy job, even if you buy genuine Chevrolet parts. It's even tougher if you buy non-genuine replacement parts. Piston weights may vary from part to part. If you replace the connecting rod, that's still another weight to consider. Compression heights also vary. When you add all of the possible ways that things can go wrong, replacing a piston and/or a connecting rod can be a lengthy undertaking which can require completely rebalancing the engine parts unless you have an accurate record of what went into the engine in the first place: connecting rod total weight, big-end weight and small-end weight, and piston type and weight (with pin).

WHAT DO YOU DO NOW THAT WE HAVE CONFUSED YOU?

If you are building a street/strip machine that is not being bored out or getting new pistons, you may now be in a dilemma. What to do? Make the pins full-floating — or live with the pressed-in pins? Change to forged pistons or live with the cast ones? Or, perhaps you are wondering how you can lower the compression ratio now that the gasoline is so bad.

If the engine is only going to see an occasional trip to the quarter-mile strip and is more designed for transportation than racing — stick with the pistons and pressed-in pins that came in your engine, especially if you are not changing the cam and valve train to allow use at higher RPM's. They'll probably go right on working perfectly without a minute's trouble. If you have an engine that is equipped with a high compression ratio, consider changing to a lower-compression cast-aluminum piston so that life will become more livable.

But, if you are building an engine which is to spend its working hours on the racing scene, then by all means disassemble it completely for a full blueprinting. Make the necessary modifications to the rods and buy the pistons needed to allow the pins to float in both. If you are buying new pistons, then consider the ones in the HD Parts List — or get some of TRW's. Most of the full-floating-pin pistons have deeper and perhaps even larger valve notches to accommodate large valves and high-lift camshafts. And, nearly all of these have 1/16-inch-thick rings which means that there will be less tendency towards ring flutter at high RPM. If you are trying to build a stock-bore or near-stock-bore 396, then remember that Chevy's stock offerings with full-floated pins are only for closed-chamber heads. And, worse yet, the rings are the 0.078-inch-thick ones that are used in the cars and trucks of low-HP types.

This head info relates to text on page 82.

SO YOU'RE "STUCK" WITH SMALL-PORT HEADS

Great! If you drive on the streets and highways, keep 'em. They are perfect for this application. Their ports are designed to keep the velocity up, even at low engine speeds, thereby guaranteeing good carburetion. Even if you are rebuilding the engine, keep the small-port heads. Have one of the porting experts we've mentioned open the chambers slightly and do a competition valve job to take full advantage of the valve sizes.

If you have a few extra bucks and are oriented towards buying new heads, get a set of the 1971 *small-port* open-chamber iron heads and add a set of the 1.88-inch exhaust valves. Add a small-port high-riser manifold with either a Q-Jet or a 3310 Holley (780 CFM with vacuum-actuated secondaries). If you are still bucks-up, add a short-duration high-lift hydraulic cam and make sure that the rear-end gear ratio is around 3.55 to 4.11 if you can afford all the gas that those will cause the engine to suck up!

A good example of what can be done with small-port heads is the 427 CID two-bolt-main engine which Bob Joehnck built for Mark Dees' businessman's Chevelle. It has unbelievable low-end torque which makes it an outstanding street performer. This car will literally spin its wheels as long as you stay on the throttle. The trick? No trick, just a set of Joehnck's highly modified *small-port* heads with a short-duration (256°) Racer Brown cam—with hydraulic lifters, of course. The manifold is an Edelbrock high-riser with a Holley 850 CFM double-pumper. Says Joehnck, "The main thing to keep in mind is that heavy valve train. Those valves are big and heavy. It's all too easy to keep your foot in it and get the engine into the RPM region where the valves will float. Buy a rev limiter and set it at the cam-maker's RPM recommendation and be sure to assemble the engine with the recommended valve-to-piston clearances and with the recommended valve springs. That rev limiter can be one of your best investments because valve float and broken valve springs have maimed, bent, destroyed and defeated more big-block Chevys than any other single problem, bar none."

The reason that this engine and others like it turn on so well is the small ports which keep the mixture velocity up at low RPM. Bob's son Fred ports and seats the heads and opens up the chambers quite a bit towards an open-chamber configuration for about $200 a pair. If you've ever done any work on cast-iron heads, you know that's a give-away price. Reworking the 1971 open-chamber small-port heads is an even better plan, according to the Joehnck father/son head-modifying team.

These tables relate to text on page 39.

Q-JET METERING JETS & RODS

Primary metering rods[1]

7034832	0.032	7034843	0.043
833	0.033	844	0.044
834	0.034	845	0.045
835	0.035	846	0.046
836	0.036	847	0.047
837	0.037	848	0.048
838	0.038	849	0.049
839	0.039	850	0.050
840	0.040	851	0.051
841	0.041	852	0.052
842	0.042		

Primary metering jets[2]

7031950	0.050	7031973	0.073
951	0.051	974	0.075
952	0.052	975	0.076
953	0.053	976	0.078
954	0.054	977	0.079
955	0.055	978	0.080
956	0.056	979	0.082
957	0.057	980	0.083
958	0.058	981	0.085
959	0.059	982	0.086
960	0.060	983	0.087
961	0.061	984	0.088
962	0.062	985	0.089
963	0.063	986	0.090
964	0.064	987	0.091
965	0.065	988	0.092
966	0.066	989	0.093
967	0.067	990	0.094
968	0.068	991	0.095
969	0.069	992	0.096
970	0.070	993	0.097
971	0.071	994	0.098
972	0.072	995	0.099

1. Diameter is largest part of rod which is in primary jet at part-throttle.
2. Diameter is jet orifice with 90° included angle at entry and 60° included angle at exit.

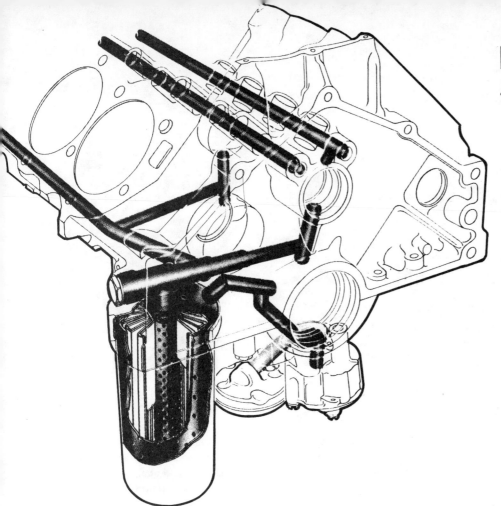

Lubrication
the stock system works for you if you leave it alone

Chevy's stock oiling system is darned near impossible to improve upon. Leave it stock and take advantage of the enormous investment in development that GM put into it in the first place.

The Stock System

Chevrolet's Turbo-Jet engine has the most troublefree and reliable lubrication system and lower end that's available in any high-performance large-displacement engine you can buy today. And, this has been accomplished without any of the desperation measures that have had to be used to make other hi-perf engines live.

Chevy's design engineers spent an enormous amount of effort in making the bottom end right! Lower-end failures in these engines are very, very rare.

This is only true if the factory's proved design is used *as is* — without the trick modifications which hundreds of experts have found to their expense to be completely unworkable. Incidentally, be sure to read the info on the types of bearings and crankshaft journals which are needed to make the big block live. If you missed that part of the chapter on building the hi-perf big block — one word tells it like it is and should be — *"stock!"*

The full-pressure lubrication system uses internal passages within the cylinder block, crankshaft and head. Oil supplied by the pump goes through a full-flow filter into a 9/16-inch main gallery along the lower left side of the block. Oil from this gallery proceeds through 5/16-inch holes to the main and camshaft bearings.

Oil from an annulus around the backside of the rear cam bearing supplies oil to the lifter or tappet galleries. Oil fed to the lifters gets to the rocker arms through hollow pushrods. 1965-66 blocks had a grooved rear cam journal and bearing to supply oil to the lifters. The '67 design is simpler, probably cheaper to produce and provides better oil control.

Spur-tooth 8-pitch oil-pump gears have 0.090-inch-wide tips to minimize leakage. Relatively high contact ratio gives smoother gear action than the previously used 7-toothers and the larger pitch diameter reduces instantaneous pressure and torque pulses which are transmitted back into the oil-pump drive shaft.

This also reduces wear on the pump gears, drive tangs and the distributor gear. The drive gear is hydrostatically load-balanced as shown in the accompanying photos. This eliminates separating hydraulic forces on the gears and reduces pump gear, body and shaft wear. Later in the chapter we tell you how to load-balance the idler gear, too.

A 5/8-inch-diameter inlet with a large suction screen produces minimum flow restrictions and eliminates pump cavitation. The limited suction capacity of previous pumps often gave pump cavitation which resulted in erratic vibration patterns. These vibrations caused erratic ignition timing (spark scatter) and increased wear on the distributor centrifugal advance mechanism.

Because everything that you do to increase power simultaneously increases heat production, don't overlook cooling the oil to eliminate heat from the engine. You can greatly improve cooling by adding an external oil cooler. Chevy has provided threaded holes for the purpose in all hi-perf blocks. You can also help the engine to rid itself of heat by improving the heat-radiating characteristics of the engine surfaces. All of this is fully explained in this chapter.

L-88 pump at left has gears slightly longer than those of center pump which is used in solid-lifter hi-perf engines. L-88 cover and body are grooved to balance pressures around drive gear. Precise fits and quality machining typify the stock Chevy pumps. Melling Hi-Volume pump, right, has longer gears than any Chevy pump strictly for selling pumps to people who don't understand the problem. These pumps are sold by many manufacturers under different trade names.

Oil Pumps

Chevrolet makes such fine oil pumps that it causes us wonderment when we see any wet-sump engine built with a non-GM pump. The high-performance pumps have 1.3-inch gears and a pressure-balanced drive gear which reduces pulsating loads on the distributor gear and reduces binding of the gear and shaft in the pump body. The pressure-balancing system also allows the oil-pump drive gear to be turned with less HP than would otherwise be required. The same pressure-balancing technique can be applied to the idler gear by adding grooves in the cover, in the walls of the pump body, and at the top of the pump (over the top of the gears when pump is installed position). This pressure-balancing also helps to avoid cavitation. The grooves which are required are shown in the accompanying photographs.

The oil pump in the big block engine is perfectly adequate, provided that you don't insist on running stupidly large clearances or switch to the wrong type of main bearings with a 360° oil groove. While high-volume pumps are available, why buy one? If they were really needed, GM could have provided them on the ZL-1 engines—but they did not. That should tell you that the stock pump will get the job done in any street machine or even an all-out racer. And, the one thing which the manufacturers of super-large pumps fail to tell you is that the bigger the pump—the more HP it takes to drive it.

Experts such as Bill Jenkins run the low-HP pump with 1.14-inch-high gears. Reason? These pumps produce plenty of pressure (55 to 65 psi), even with 20 or 30-weight oil. The "small" pumps can be modified to add the pressure-balance feature. So use the stock oil pump that came in your engine. Add those grooves for the pressure-balancing scheme as required and make sure that the pressure relief passages are clean and that the pressure relief valve operates. Don't shim the spring. Get the end clearance down to 0.0025. That's what it takes.

In a POPULAR HOTRODDING article, "The Hot Setup for Super Lubrication," Bill Jenkins said, "You have to keep in mind that the bigger the pump, the more oil it takes to feed it, and the more oil you have to get through the bypass when it blows the valve open—and the more chance you have for cavitation."

Sharp tuners will wonder why the small-block Chevy oil pump can't be used in the big block. Actually, you could use one, but you'd be defeating the engineers' work. They purposely built the big-block pumps with more teeth on the gears to reduce pulsations transmitted to the distributor and distributor-drive gears. Previous pumps (with fewer gear teeth) caused gear breakage and spark scatter at high RPM, hence the changeover to the "toothier" design in all big-block pumps. You'll want to modify your pump so that it is fully pressure-balanced as shown in the photos.

Part of the reason that so many heavy-duty oil pumps are sold can be traced to mechanics who won't believe that the 0.0025-inch main and rod bearing clearance is adequate. So, they open up the clearances.

The end result is that the stock pump can no longer keep up with all of the extra leakage that these "experts" have built into their engines. The big-block engine thrives on factory-recommended clearances. Fortunes were expended by GM to make the engine right. So, unless you have a couple of fortunes to throw away, stick with the recommendations which were developed as a result of massive engineering efforts.

Perhaps the most important thing which must be set correctly in the oil pump is the end clearance. This should be in the neighborhood of 0.0025-inch. If it is more, then the pump will not prime as easily and will not be as effective. If the end play is too great, the clearance can be reduced by sanding the pump body on solvent-wetted No. 220 Wet-or-Dry paper supported on a thick piece of glass. If the clearance is too small, the gears can be sanded on one end the same way. Clearance can be checked with a depth micrometer, Plasti-Gage, or with feeler gages. Stock pumps which we measured had 0.003 to 0.004-inch clearance.

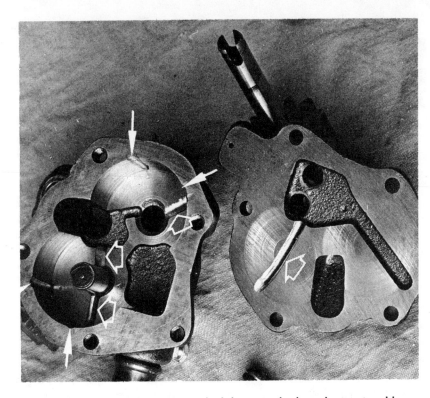

Outline arrows show where Bill Jenkins reworked the pump body and cover to add pressure-balance grooves across idler gear at each end of pump to balance that gear hydrostatically. The factory balances the drive gear on hi-perf pumps, but you can balance both gears in any big-block pump. White arrows indicate grooves added to the vertical sides of the gear bores to assist in the balancing act. Popular Hotrodding photo.

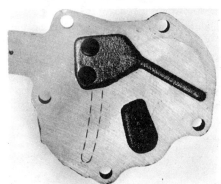

Dotted lines show where you should add grooves to your stock pump to get full load-balancing benefits for both the drive and idler gears. Do it!

Lightly chamfering or deburring the gear teeth edges with a fine file is recommended practice in getting any pump ready to run at high RPM.

End clearance should be checked after the gears have been deburred at both ends. This oil-pump body had .004" clearance—will be sanded down at the joint to reduce the end clearance to recommended 0.0025 inch.

113

The pickup tube must be brazed or tack-welded to the pump body. A pickup tube shaking loose and falling out of the pump is not uncommon. The bottom of the pickup should be spaced about 3/8-inch from the bottom of the pan.

Pickup inlet is extended to fit into bottom of deep-sump pan. This is from a Bill Jenkins engine. Note that pickup opening faces rear to pick up oil during hard acceleration. Locking capscrew (arrow) and Loctite secure pickup tube in housing, instead of welding the tube to the pump body, as is common practice. Photo courtesy Popular Hotrodding.

The pump attaches to the rear main bearing cap, so it is important to make sure that the hole in the pump aligns with the hole in the cap. Some chamfering may be needed to get the holes lined up with each other. A cap screw goes through the hole, so remember that the oil has to pass through this same area and exit on the opposite side of the cap screw.

Almost all racing engine builders weld the pickup to the pump, although we noted that Bill Jenkins adds a lock-nutted set screw to hold the pickup in the pump body. The high-performance pumps come from the factory with the pickup tack-welded to the pump body in several places. These few weld spots should not be relied on to do the job in an engine which will be raced. The stamped-metal shield over the screen in the bottom of the pickup must be left in place because it helps prevent air from being picked up with the oil. It is recommended that the pickup be repositioned for the particular use the engine will see. For oval-track racing, the pickup should be relocated on the right side of the pan. For drag racing, the pickup can be relocated to the extreme rear of the pan so that the force generated by the acceleration of the car will keep oil around the pickup. With such a setup, a vertical baffle must be placed ahead of the pickup to avoid oil starvation on deceleration.

Regardless of the position in the pan, the pickup should be spaced 1/4 to 3/8-inch above the pan bottom. If an oil cooler is to be used, it may be necessary to shim the oil pump spring 1/8-inch or install a stiffer spring to restore the desired 50 psi hot oil pressure which should be run in a high-performance Chevrolet engine. If the spring is shimmed, place the shims inside of the pressure bypass valve rather than at the other end of the spring.

Don't try to run too much pressure. When the oil is at operating temperature, pressure should never exceed 55-60 pounds. Anything over this just whips more air into the system. Another way of putting this is: you can't aid bearing life with more oil pressure.

Surge gates inside a modified oil pan allow oil to move toward the pick up, but not away from it.

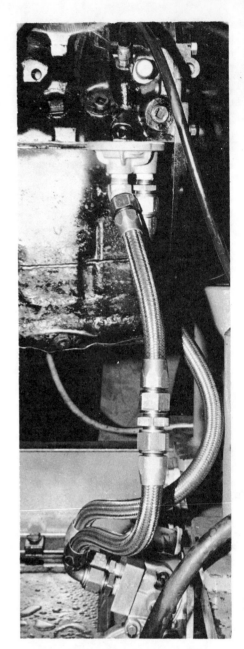

On the dyno or off, aircraft-type steel-braided hose is the hot ticket for fail-safe plumbing of oil or gas lines. Various sizes of lines and scores of combinations of fittings can be purchased at surplus stores or from race car supply firms.

You have seen advertisements for swinging pickups which follow the body of oil during acceleration or hard braking. Swivelling type oil pickups still need development to get full potential. At the present time they are largely used in funny-cars and in road-race machinery. When used in road-race machines, the pickups have strange quirks which you might not recognize or think about until too late. You don't need or want one for your street/strip machine.

The pickups work fine if there are sufficient "g" forces. If you declutch and shift, the pump will gulp air and drop oil pressure. If you "float" over a hill and are slowing down, the same thing occurs. Mechanics with whom we talked claimed swinging pickups can cause failures if the pickup happens to stick in the forward position just as you get back on the throttle to accelerate hard. Perhaps a pickup which swings 180 degrees from side-to-side only will be best because it will take care of cornering loads. Such a pickup could be used with baffles to take care of the acceleration and braking forces. Baffling the pan is a better solution for oil control. If you do choose to use a swinging pickup, for whatever reason, then the swinging joint should be pressurized with filtered oil through a separate small filter. If this is overlooked, chances are good that a tiny piece of junk will wedge itself in the joint and prevent the pickup from swinging. You can figure out where that will leave your engine. The ultimate design will probably end up with ball bearings supporting the pickup, but presently available designs usually manage to suck air through the ball bearing.

This is the oil pickup on Wally Booth's Pro Stock engine which puts out something in excess of 650 horsepower. You'll notice that the pickup is stock.

Out of the box from Chevrolet comes their heavy-duty deep-sump oil pan 3977591 which holds six quarts. With the windage tray in place, it does a pretty good job of keeping the oil near the pump pickup and away from the crankshaft.

Oil Pans

Milodon Engineering of Chatsworth, California makes a number of different pickups and extensions for pumps so that the pickup body always remains immersed. Milodon, Avaid, Ron Butler Racing, and Tri-Metric Engineering are a few of the firms making a wide variety of performance-oriented oil pans for the big block. An array of baffles, swinging gates, and sump sizes and shapes are available for specific applications. Depending on the complexity of the pan required to do the job, these special pans will cost a few hundred dollars.

A low-budget high-performance trick is to use the semi-circular Corvette tray baffle 3967854. Keep in mind that the Corvette engines have a five-quart oil capacity oil pan with a trap-door baffle. Optional oil pan 3977591 hold six quarts in a deeper sump and restrains oil movements by efficient baffling. Be sure to use the correct pump with this pan. It has a longer inlet tube to put the screen at the bottom of the deeper pan. Many automotive-accessory firms make and sell magnetic drain plugs. Use one.

Extended tube puts the pickup next to the bottom of deepened sump...to keep pickup in oil when car is accelerated hard — or braked! Stock pump/sump are shown at bottom for comparison. These are Bill Jenkins' parts. Photo courtesy Popular Hotrodding.

Weiand front cover hides a gear drive. Hilborn fuel pump will feed injectors. Degreed damper snugs onto a polished high-performance crank which spins above a fully baffled and gated Aviad pan. This big block is going racing—hopefully to earn the money that it took to build this high-pressure thumper.

Oil drainback into pan is through troughs as indicated by arrows. Details of the need for this intricate stock baffling are contained in the column of text at right.

Stock Pan Baffling

Baffles in the stock pan are there for very definite reasons — to save the engine when the oil gets low and you have your foot into the throttle. They also keep the oil pickup submerged during acceleration or deceleration. The original Society of Automotive Engineers paper, "Chevrolet Turbo-Jet Engine," by Richard L. Keinath, Herbert G. Sood and William L. Polkinghorne, had this to say about the pan baffling. "A lubrication problem showed up early in the design stage. Chassis and ground-clearance requirements limited the oil sump to a shape with a capacity of four quarts. It was found that a high-performance Turbo-Jet version, with maximum bearing clearances throughout, would experience oil starvation at over 5,000 RPM when the oil level was at the "ADD" line or with three quarts in the sump.

"The oil was pooling ahead of the forward slosh baffle, which is used to keep oil in the sump during rapid vehicle deceleration. The metering hole in the baffle was enlarged to permit enough oil drain back and avoid starvation. However, this resulted in poor pressure during panic brake stops.

"Drain back to the sump was significantly improved and starvation eliminated by the addition of two troughs in the oil-pan slosh baffle. The troughs permitted the oil to flow rearward into the sump without restriction and the baffle trapped it during panic stops."

According to the same paper, the starvation zone at the ADD oil level was raised from 5,000 to 6,500 RPM. This would seem to indicate that it will be a good idea to keep the oil at the FULL mark if you have the four-quart sump. It should also caution you against trying to redo the baffling. You might end up worsening whatever condition it is that you think needs correcting.

Breathers are not placed on valve covers solely for looks. Pressure in the valve-train cavity and crankcase rises as RPM goes up. Seals and gaskets have been known to blow out if there are no vents. This is not the recommended way to vent the big Chev's.

Tray atop injector stacks is a good idea. So are the collectored headers. Boat builders seem to understand that straight stacks that give away HP don't make much sense. But three breathers on each rocker cover? That's like bandaging your elbow when your knee is skinned! The numerous breathers were required because the stock covers with their efficient oil separators were removed. Stock covers also incorporate drip rails which ensure that the rocker balls are correctly lubricated by oil throwoff. Not all aluminum covers have this feature. Special covers sometimes cause more problems than you'd want.

Engine breathing and venting

Nearly all mechanics start out with the idea that the stock breathing for the big block is inadequate. Nothing could be further from the truth. The stock covers are equipped with oil separators which are designed to prevent oil pullover at air flow up to 10 cubic feet per minute—which is a lot of air flow. You can use these rocker cover vents to make a system adequate for any racing engine.

Just connect the two rocker covers by 3/4-inch ID or larger tubing which runs up and across the front of the engine with two or more breather-type oil caps in the piece of tubing which is at the highest point. If you braze elbows into the stock covers, be sure to leave the stock oil separators intact. When you install this type of system, you eliminate any need for expensive and ineffective hotrod breathers which usually blow oil all over your engine. The system which we have described was first used on the Dodge/Plymouth Trans Am entries in 1970 and was subsequently used on those monster engines that run in the NASCAR events. Success in both of these applications has proved that this is the system to consider.

There are cast-aluminum covers—and even cast-magnesium rocker covers available for the big block—but most do not include any oil separators and special breathers must be constructed to bring back the crankcase pressure relief which was previously provided in the stock covers. This may be one instance where the cosmetic benefits of the special parts should be foregone in favor of the real benefits and built-in breathing of the stock parts.

If an electric fuel pump is used, the stock fuel-pump mount can be used for venting hardware.

Why all this fuss about venting? The engine is nothing more than an air pump, which means that the same volume of air is moved around in the bottom of the engine as is inhaled into and exhaled out of the combustion chamber. Couple this with the air and oil mist whipped around by that massive rotating crankshaft and you quickly see that the bottom side of any large engine gets to be a violently active area when it comes to air velocity and direction. The engine simply needs a plenum chamber at the bottom end where all of the vented air from the downstrokes can be displaced through vents. High-RPM ring flutter is often directly traceable to pressure at the *lower* end—which correct venting helps to reduce.

The right venting also helps the engine to run cooler.

Two breathers on tube connecting to the stock rocker covers will do the job.

Windage Tray

The semi-circular tray baffle 3879640 under the crankshaft drains the oil away from the crank and at the same time prevents the crank throws from hitting an accumulation of oil. This baffle should be installed on any large-block Chevrolet engine—regardless of the type of oiling system used. It's worth several HP at high RPM's.

Oil Pressure Gage

Use a direct-reading oil-pressure gage with at least 3/16-inch ID line for rapid gage response. In this regard you should know that the majority of engine-bearing failures are a direct result of a severe drop in oil pressure when the oil pump picks up air while the car is in a high-speed turn. In a situation such as this, the driver is so busy that he does not notice the gage. A responsive gage mounted near the driver's line of vision is a must for a high-performance vehicle, but all of this is only as good as the driver's willingness to observe the gage.

One way to circumvent this problem is to install a Mallory "Sonalert" warning horn in a circuit with an idiot-light pressure sender. The vehicle battery can be used or a tiny Dura-Cell battery can be used. The 80-decibel blast—that's real loud—which this tiny (under 2 oz.) warning horn produces should wake your driver to the fact that he should get off the throttle until oil pressure returns.

Still another remedy is to install an engine shut off switch which turns off the ignition when oil pressure falls below a safe level. Ansen makes such a device.

If oil pressure does drop there are several areas to start exploring to locate the cause. Simply adding another quart of oil will sometimes cure the problem. Higher engine flow rates can also cause an oil void in the sump—thus a drop in pressure. This is often caused by excessive bearing clearances and higher than necessary oil pressure.

Incorrect oil-pan baffling (usually homegrown over-baffling) which prevents the engine oil from draining back into the oil pan while the car is in a turn or is accelerating can also cause problems.

Idiot Light

If your car only has the idiot light oil-pressure warning, consider that this is far too slow and unreliable for any high-performance activity. When your oil-pressure light goes on, this only tells you that the oil pressure is somewhere between 2 and 6 psi! Or, you could be down to zero pressure. By the time that the light comes on, damage has probably already occurred. For this reason an oil-pressure gage should be considered essential. However, you can keep the idiot-light circuit and sender intact and use it for a catastrophe warning by wiring it to a large clearance light on the dash or to a horn relay so that you'll either be blinded by the big light or awakened by the horn blowing in the event of oil-pressure loss.

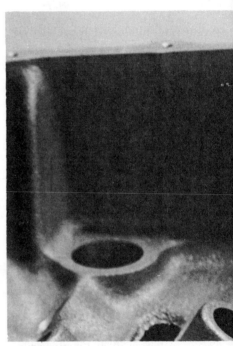

All later large- and small-block Chevys (mid-year change, 1969) have no distributor-housing boss extended from the lifter valley floor to the uppermost portion of the block. Chevrolet decided the long boss was not necessary and thus eliminated that portion of the casting..

Two of the semi-circular windage trays can be welded together to make a longer unit which fits under all of the rods. Plastic-insert nuts (self-locking) should be used to attach tray to main studs. Use flat washers under the nuts to distribute clamping forces or tray may crack from flexing. These trays were discontinued at the end of 1970, so you may have to use two of the flat windage trays or fabricate your own. Obviously, the long windage tray will not work with those pans which have minimum oil capacity for chassis clearance reasons.

What kind of oil?

Chevrolet has even provided an oil specification for the big block. They recommend 30 to 50 weight aircraft or other ashless high-performance oil such as Valvoline Racing Oil, Gulf Formula G Oil, or Sunoco 10/50 or 20/50 Multigrade. Most of the oils that you purchase at the discount store or corner service station are low-ash oils. The ash which remains, although small in amount, creates preignition-causing deposits. Many automotive oils were once ashless but have gone to the low-ash formulation as a result of incorporating anti-rusting agents and other agents re-required to make the oils—and engines—live with the attitudes prevalent today. However, you should realize that the oil companies cannot protect your engine when you start using oils without some of the long-life additives. You'll have to make sure that you change the oil and filter frequently—something that most people have fallen out of the habit of doing in recent years.

DO NOT RUN HIGH VISCOSITY OILS. Stay on the lighter side of Chevrolet's recommendation. There is no need for the heavy stuff which merely complicates the oiling, perhaps creating bearing problems which you'll probably blame on something or someone else. Use the right grade of oil—usually 30 weight—of the recommended type and stick with it. High-viscosity oils do such terrible things as tearing up filter cans, not flowing to the bearings when the engine is cold, etc. Above all, stay off of the throttle until the engine is fully warmed up, no matter how cute the chick that wants to hear your engine go "rump-rump."

Bob Joehnck says, "Most large-block failures are due to racing with too heavy an oil. Some drivers insist on 'jazzing' the throttle before the engine is thoroughly warmed up. Bearing failures inevitably result. We run 30-weight oil in all of our engines—especially in the wintertime—making sure that this relatively 'thin' oil is thoroughly warmed up before ever revving the engine. Oil must be warm for these engines to function correctly. Some of the best-looking lower ends that we have seen have been in engines that have just finished a 200-mile race at Riverside where it is 110° in the shade. Naturally, we run coolers so that the oil temperature never exceeds 225°, but oil which is too cold cannot be recommended for these engines. When the oil is too cold, the pump can't move it regardless of the clearances that you have used. The Harrison NASCAR cooler is an absolute essential for the big-block which is being road raced ... so don't neglect to install one—or you'll wish that you had."

As you peruse the test information which we obtained from Chevrolet, you'll note that all of their tests were accomplished with 30-weight ashless oil. ALL such testing, including durability tests, was done with this type of oil. None heavier was ever required or used. Why, then, do so many drag racers insist on using 50-weight oil? Because this is the only weight that they can use which will give any oil pressure when the clearances are opened up unnecessarily. If you are getting excessive oil dilution, check why and fix the problem. When stock lower ends have the reputation of being hell-for-strong in stock configuration — and reliable as a locomotive — why not start off with a stock lower end and 30-weight oil? If the engine does not stand up to your kind of racing, maybe you didn't put the engine together right.

Arrow indicates takeoff for 3/16-inch oil-pressure-gage line. Fitting toward front of this pro-stocker's block is used for preheating the engine with hot water circulated from the engine of the tow car. Stahl Associates offers the quick-disconnect system. Note that the engine is painted black to help radiate heat.

Oil Additives

Only one oil additive should ever be considered for use in the big-block Chevrolet and that is GM's Engine Oil Supplement (E.O.S.) which can be used on the cam and lifters — and on the rocker-arm balls — when the engine is assembled. But, don't use this additive in your engine on a continuing basis if you are racing it. Any additives which have calcium or barium in them will cause preignition — and E.O.S. falls into this category.

Additives ultimately cause preignition in a racing engine because detonation in the end-gas area of the combustion chamber heats the barium and calcium deposits until they are glowing, thereby starting the preignition cycle which will completely destroy the engine if allowed to continue.

E.O.S. is available at GM dealers in 16 & 32 oz. cans. It is probably the best lubricant-additive that you can use in any new engine. With all of GM's research behind it —you can be sure you're getting your money's worth! Some cam grinders even include a can of it with every cam and lifter kit which they sell because the stuff really helps to avoid lifter scuffing and cam-lobe wear during those critical first few minutes of operation.

Avoid other oil additives because they may not be compatible with the additives which the oil manufacturer has used.

Oil-filter lineup: On left is canister-type filter with adapters to install the unit into big blocks. '65-67 blocks take adapter 5573979 (1), '68 and up police and truck engine adapter 3951626 can be used on any '68 and later engine (2). Canister is 5574535, cartridge is 5576054. Screw-on filters use adapter 3952301 on '68 and later blocks (3). Standard short filter 6438261 can be replaced with heavy-duty model 6438384 (PF-35). Canister filter has more filtering area and is cheaper to replace. Using the screw-on-type filter with the 1965-67 engines requires using adapter kits as made by Eelco or Mr. Gasket.

Oil Filters

Chevy has put the oil filter right on the block so there's no need to arrange for a remotely mounted filter at big expense. Some builders think that the screw-on filters introduced on 1968 engines and used on all big blocks built since that time are necessarily better than the older canister-style filters. This is not the case, so don't rush to buy an adapter kit to put the late-style screw-on filter on your pre-1968 engine. All that the later filters "buy you" is higher replacement cost. Don't complain when you are cleaning out the messy canister—think about the money you are saving. And, you can take some pleasure in knowing that the larger canister-type filter has more filtering surface to clean your oil. In fact, some racing-engine builders won't use the screw-on filters. They convert the 1968-and-later blocks to use the early-style filter. These guys order a kit from Chevy, P/N 5574538, throw away the adapter that's in the kit—or swap it for adapter 3951626.

Another way to get extra capacity is to replace the standard screw-on filter with the longer, heavy-duty version.

The filter pressure-relief valve in the filter adapter is a disaster preventer for folks who refuse to change their filters and oil regularly. The valve should be plugged so that all oil is filtered before it can get to the engine. Watch that oil-pressure gage that you added. When the pressure starts to drop (assuming you are measuring the oil pressure *after* the filter), replace the filter cartridge because it's obviously become too clogged to work any more.

The first time that you fire up your engine, run it for 20 or 30 minutes at a fast idle. Drain the oil *while it is still warm* and promptly change both the oil and the filter cartridge. New engines are always full of lint and other trash — much of it from shop rags. This lint is the very worst culprit for plugging up filters. The price of the oil and filter that you throw away is cheap insurance against losing an engine. If you have plugged the filter relief, then the new filter ensures that you will have good oil pressure. If you have left the relief valve stock, dirty oil will bypass the filter whenever the oil is cold or the cartridge is plugged — and it will go directly to your engine's bearings and lifters.

Of course, when the filter relief is plugged, you have to avoid jazzing the throttle when the engine is cold. You will have to let the engine warm up before running any high RPM's or the pump by-pass may not be able to cope with the pressure relief job — and your filter could be burst open by the high pressures which can occur with a cold engine.

So, what's so great about a full-flow filter and why should you continue to use the one that Chevy builds into their engines? Let's look at some facts and figures published by a competitor, the Ford Motor Company, in an SAE paper published in the 1950's. A full-flow filter reduced engine-component wear by the following amounts: 50% in crankshaft wear, 66% in wrist-pin wear, 19% in cylinder-wall wear and 52% in ring wear.

Much of the engine wear occurs from dirt or particles of metal which are seldom completely removed from the engine when it is first built or overhauled. Some of the engine wear comes from wear products—bearing particles, pieces off of the camshaft or lifters, minute bits of carbon which somehow manage to get past the rings, and tiny chunks of aluminum which are worn off of aluminum heads or valve-spring retainers by the valve-springs. This is especially true when the valve-springs are seated directly against the aluminum surface of the head—instead of against shims as is correct procedure. Of course, if you stick with steel retainers as supplied by Chevrolet, you'll eliminate one possible source of aluminum chips in your oil. If you have iron heads, these, too, should be equipped with steel shims at the bottom of the spring seats.

When these chips are recirculated through the engine with the oil, additional wear is caused, adding to the quantity of unwanted junk in the oil on a continually increasing basis so that the oil becomes a carrier of wear-producing material instead of a flow of life-sustaining lubricant.

Oil Cooler

An oil cooler can be connected through hoses to the high-perf blocks by removing the plugs above the oil filter pad. You have a choice of what to install in the rear hole. Either plug the hole as the professional racing engine builders do, or follow Chevrolet's published "safe" recommendation and install a six-cylinder oil-filter by-pass valve 5575416. Either will cause oil to flow through the cooler, but the plug will ensure that *all* oil goes through the filter instead of straight to the engine.

If you use the bypass, oil temperature may not be reduced by your cooler because the oil may choose the path of least resistance and take the "shortcut" into the engine via the bypass.

When installing your oil-cooler plumbing, keep all of the hoses at least 1/2-inch ID and avoid elbows or reducing fittings wherever possible. One 90° elbow is equivalent to 10 feet of pipe of the same diameter—so you can see how elbows can restrict the flow. Small hose also restricts flow through the cooler.

Keep the hoses short. Use hoses designed for hot-oil service. They must be specified for long life at temperatures to $350^{\circ}F$ or so. Aero-Quip steel-braid-covered hose is possibly the best you can buy for remote oil cooler or dry-sump plumbing. This hose or non-sheathed hose for oil service can be found at firms specializing in truck or aircraft parts.

The oil-cooler hoses should be filled as they are installed. Or, if this is not possible, the pump must be turned with an electric drill until the entire system is filled. Failure to do this could cause bearing damage because the engine will have to turn over a lot of times before the bearings get pressure.

A big Harrison cooler, 3157804, is listed in the heavy-duty parts list. But it is not ready to use as it comes out of the box. You'll have to take it to a heliarc expert to get it cut apart and modified so that the oil flows into one end of the cooler and out the other.

Low-restriction coolers with the inlet/outlet on opposite ends are available readymade, so unless you get your heliarc work done dirt-cheap, you may want to start off with a non-GM part that needs no modification. A Boss Ford cooler made by Karmazin Products is one possibility, Ford P/N C90Z-6A642-A with mounting brackets C9ZZ-6B633-A and C9ZZ-6B634-A. The brackets are an absolute necessity, so don't overlook these rubber-shockmounted pieces when you place your order. The cooler is all-steel, so it weighs a bit more than the Harrison. It is 7" tall, 15.6 inches wide and about 2" thick. Ford high-performance engineer Jeff Quick told us that the cooler can be relied upon for about a $30^{\circ}F$ temperature drop between the inlet and outlet, depending on inlet temperature, of course. He also cautioned us that the cooler should be mounted where it gets fresh air ducted to it with a free opening behind the fins. The cooler cannot be expected to work if it is mounted against a flat surface—or immediately ahead of the stock radiator. Either type of mounting will impede the flow of air through the cooler, which is what's essential in this instance. Mount the cooler so that it is out of the way of rocks and dirt which could clog the fins and reduce the unit's cooling capabilities.

Lest the over-$100 price of either the Harrison or Karmazin cooler tempt you to try to use an automotive heater radiator core for a cooler, let us assure you that such items are not designed to withstand oil-pump pressure. The pressures developed in cooling systems are far less than those in an oil system.

A lot of air-conditioning evaporators are seeing racecar duty as oil coolers. They are available in a wide variety of sizes and shapes and they're reasonably priced in a scrap yard. Select one with a tube diameter of close to half inch and have fittings heliarced on to accept the oil lines. Have the unit pressure checked. As a rule, an evaporator will withstand more than 100 PSI which is adequate for cooling big-block oil.

A common mistake when modifying an oiling system on a high-performance engine is over-cooling the oil. Oil should be kept at about $200^{\circ}F$ to flow properly. Oil temperature should be measured where the oil comes back to the engine from the cooler. However, it is important to install the temperature sender so that it does not restrict oil flow in any way. A temperature range of 180 to $230^{\circ}F$ is normal. Oil temperature should never exceed $230^{\circ}F$—but often will, regardless of how hard you try to keep it at

Harrison cooler listed in HD Parts List is a good cooler if you modify it internally so that oil flows from one end to the other. The inlet must be at one end and the outlet at the other.

a cooler temperature. Chevy suggests that the oil in the pan should never exceed $300^{\circ}F$.

Use your engine as an oil cooler

The professional racers who paint their engine blocks, rocker covers and oil pans flat black are doing this for reasons with a sound basis in engineering fact. P. E. Irving, in his excellent book MOTORCYCLE ENGINEERING (available from Autobooks in Burbank, California) has this to say on page 183, "A polished surface emits less heat by radiation than a black one. Rate of heat emission from a polished surface is approximately one-tenth that from the same surface covered with a thin film of lampblack, and the emissivity of a cast-aluminum surface is increased about 10% by a thin coating of black paint." You only have to look as far as your stove to notice that tea pots are polished to keep heat in—so why do the same to an engine?

Thus, a black surface is *ten times as efficient* as a polished one. ALCOA's Engineering Handbook indicates that Irving's comment is quite conservative. ALCOA compares an as-cast surface with

121

one which has been black-anodized to a depth of 1.7 thousandths. The black surface is *more than ten times better in heat-radiating ability than a plain cast surface.* Remember these facts when you are tempted to start polishing and chroming various engine parts which can contribute to cooling efficiency.

Another noted mechanical engineer, Mr. Julius Mackerle, in his book, AIR-COOLED MOTOR ENGINES (also available from Autobooks), states that it is an error to assume that using a greater quantity of oil will reduce oil temperature. Every ad for a big sump and almost every magazine article which discusses why racers install big sumps tells you that the use of larger quantities of oil reduces temperature. This common misunderstanding is completely untrue and utterly ridiculous! *Additional oil capacity does not lower oil temperature*—it merely increases the time required for the oil to attain a stable operating temperature: not usually of any real importance. Oil changes cost slightly more and the engine requires longer to cool off after it has been run. Additional oil capacity is helpful in a long race, providing the pickup is moved to the bottom of the bolt-on sump to make all of the oil available for use and so that the pump always gets the coolest oil. Additional capacity is insurance against losing an engine in the event that unexpected oil-consumption problems develop in a racing situation.

Mackerle further remarks that "A finned sump does not aid cooling to any great extent as the oil does not flow down the cooled sump walls. Cooling is more intense on the crankcase walls, over which the oil flows in a *thin film* . . . Best oil cooling is obtained by a tube-type radiator . . ."

The reader should take note of the words "*thin film*" as these are the key to understanding the removal of heat from oil.

If you use the stock rocker covers, don't use the chrome-plated ones! Additional heat-removal capability can be added by welding or brazing sheet-metal fingers and/or baffles to extend into the hot oil to transmit the heat to the radiating surface. The orange-enamelled covers should be refinished with a *thin* coat of flat-black paint. The stock covers have a lot of surface area which you can use advantageously in helping your engine to rid itself of destructive heat.

If you have already purchased special aluminum rocker covers or chrome-plated ones, or if you have a chrome-plated pan, you can still use these parts to advantage. Either sell them to get your money back out of them so that you can buy new stock parts, or "de-polish" them by sand-blasting the polished surfaces. Black anodize the aluminum parts. Black anodizing is a plating process which many plating shops can provide for you at reasonable cost. Incidentally, the black obtained by anodizing may turn out somewhat spotty because castings do not usually anodize perfectly. This will not impair the heat-removal characteristics.

Parts which are not aluminum, or aluminum parts which are not being black anodized, can be painted with a thin coat of self-etching flat-black paint of the non-insulating variety. The entire block can also be painted on the outside with flat-black paint. This is one place where you don't want Sperex's fine insulating paint. You want to let the heat out—not keep it in. You can paint the headers or exhaust manifolds with Sperex paint so that these will not add as much exhaust heat to the engine compartment, thereby allowing the other parts to cool off more rapidly.

Although seemingly insignificant, these efforts will reward you with a cooler-running, longer-lasting engine.

You might think that some fins welded or brazed onto the bottom of the pan could assist the engine in getting rid of heat. Let's see what Phil Irving has to say about this on page 239 in the book already mentioned. "While oil is good at collecting heat, it is very bad at getting rid of it again, because the layer directly in contact with a cool surface increases its viscosity and stays there, acting as an insulator and effectively preventing heat being dissipated from the hotter oil in the interior. Ribbing a sump which contains a quantity of oil is not very effective unless these are internal ribs also to transfer as much heat as possible from the body of the oil, but ribs placed on areas against which hot oil is violently thrown by centrifugal action can be made to radiate a lot of heat." From this we can see that ribs, if you add them to the pan, should really be added to the sides of the pan where the oil is flung against the walls by the crank.

Dry Sump Systems

Drag boats, ski boats, and a wide variety of road-racing vehicles can benefit from a dry-sump system instead of relying on the conventional single pressure pump.

The majority of internal combustion engines produced in the U.S.A. have "wet-sump" oiling systems. A wet-sump design carries the engine oil in a reservoir (sump or pan) directly below the engine crankshaft. An oil pump in the pan forces the oil to all engine bearings, rocker arms, etc. Gravity returns the oil to the pan for recirculation. The wet-sump system is impractical when racing with low chassis height and reduced ground clearance. These factors won't permit using a deep oil sump with an adequate supply of oil. A shallow sump produces oil foaming and excessive windage. At high RPM, the crankshaft can keep enough oil airborne by "windage" that 25 to 50 lbs. ft. of engine torque is lost.

Dry-sump systems have been used successfully in aircraft engines for many years. "Dry-sump" design uses a remote engine-oil reservoir, generally level or above the pump system so that gravity ensures priming. Oil is pumped from the reservoir to the engine bearings, falls by gravity to the "dry" sump and is returned to the remote oil reservoir by a separate pump system. In a dry-sump system, oil capacity is limited only by the size of the remote reservoir: two to three gallons is a common size. Auxiliary oil coolers and filters are readily accommodated.

Chevrolet introduced a three-geared pump for a pressure/scavenge dry-sump system, but dropped it when the Weaver Brothers pumps became available. The Weaver pumps have become the racers' standard for dry-sump-oiling systems. It allows holding oil-pan depth to a minimum—always less than flywheel depth. Two separate oil pick-ups should be used: one for the front of the oil pan and one for the rear. Due to pickup location, acceleration, braking, windage or turns do not interfere with getting all the oil out of the sump so that it can be returned to the remote reservoir. The stock oil pump is eliminated and replaced by an externally mounted, gear-belt-driven pump consisting of at least three totally independent gear pumps. Two gear pumps are separately connected to the front and rear oil-pan pick-ups and pump oil back to the reservoir. The third gear pump delivers oil from the remote reservoir to the engine bearings through a filter and cooler. This pump is equipped with an externally adjustable pressure regulator. Aircraft-quality hoses, fittings and plumbing practice must be used throughout with oil filters and coolers as required.

If your racing machinery requires a dry-sump system, expect to spend well over $1000 for a simple dry-sump installation—more for a complex one with extra scavenge stages, injector drives, etc. Weaver Brothers also provide cog-belt drives for installations where only a water pump, or a water pump and an injector pump, are used.

For a steel pan with built-in screened pickups, you'll want to contact Ron Butler of Butler Racing in Culver City, John Mason of Mason Eng., or Ed Pink of Ed Pink Racing in Van Nuys. These are all California firms.

The following tips should assist you when installing a dry-sump system.
1. Eliminate the engine oil pump completely and plug the rear oil-cooler holes in the block and rear bypass valve position.
2. The scavenge pump/s should have three times the capacity of the pressure pump.
3. At least two scavenge stages should scavenge the oil pan and one stage should connect to the rear outside of the rocker cover on the predominant outboard side of the car. This depends on the course and whether it is run clockwise or counter-clockwise.
4. Do not run scavenged oil through the engine oil cooler but return it directly from scavenge pumps to supply tank through -12 or 3/4" line. Install a coarse-screen aircraft filter in this line to keep contaminants out of the pressure pump and pressure-bypass valve.
5. Use a -12 or 3/4" inlet line to the pressure pump from the supply tank.

Weaver Brothers' dry-sump pumps are the racers' standard. Pump shown here is a three-stager (two scavenge, one pressure). Two stages are visible ahead of the mounting bracket. Drive is by toothed belt. Dry-sump pan makers are mentioned in the text.

6. Pass oil from pressure pump through engine oil cooler and remote oil filter and then into the forward oil cooler hole in the block. The oil filter on the engine can be used if you have an aluminum block. There's no way to connect the filter in an iron block unless some added drilling and tapping is done—but this would probably be cheaper than installing a remote filter. Make every effort to reduce restriction in the oil-cooler circuit. Do not connect oil coolers in series. If more than one oil cooler is used, connect them in parallel, i.e., tee the oil line and pass the oil into and out of both coolers simultaneously.
7. Do not over-cool the oil. Racing oil must be at about 200°F to flow properly. Measure oil temperature between the oil cooler and the engine. Keep it between 180° and 240°F when thoroughly warmed up.
8. Do not exceed 55-60 psi oil pressure (hot) because excessive pressure aggravates oil aeration and creates scavenging problems. Oil pressure over 55 psi is not necessary for good bearing life.
9. Run a full-length semi-circular tray baffle under the crankshaft with louvers to draw the oil away from the crank.
10. Design the oil supply tank as tall and as small in diameter as space permits. It should hold a minimum of 8 quarts of oil with enough air space above the oil to allow oil-air separation. The tank bottom must be level with or slightly higher than the pump inlet so that gravity will prime the pump.
11. Build the engine with the correct lifters, rocker arms, rear cam bearing and clearances to require a minimum of oil flow. Such attention to detail is the greatest asset to a correctly functioning dry-sump system.
12. Vent both the engine and the supply tank, or vent the engine to a correctly vented supply tank. Keep vent lines of adequate size (one -12 or two -10 lines) to keep from causing any pressure build-up in the crankcase. Making these vent lines too small is a common mistake. Existing breather holes in the engine rocker covers are an excellent place to vent from. Leave the production oil separators in the rocker covers under the vent holes. These are designed to handle 10 CFM without oil pull-over—provided that the vent lines are of adequate size—out of the system. You'd be hard pressed to design something with this kind of oil-separation capability and GM gives it to you in the factory-direct low-cost parts.

This material relates to text on pp 51 and 158.

Tachometers

An accurate, mechanically driven tachometer should be used on any big-block Chevrolet used for serious competition. The Jones Motrola tachometer is universally accepted as the best one. The higher priced unit includes a "tell-tale" needle. Chevy engineer Bill Howell says that some Can Am engine builders have had disastrous experiences when mechanical tachs were found reading *1000 RPM slow* at 6000 RPM! Even a few-hundred RPM difference would be risky. He recommends checking mechanical tach (or any tach for that matter) accuracy against a known speed device several times during the year and *always* before installing a tach for the first time.

Electronic tachometers are fine for street use, but few professional racers will trust them for competition, with a possible exception which we've noted in the section on rev limiters.

Rev Limiters

Recommendations from Chevrolet are clear and plain: If the valve train is matched to the cam being used and there is adequate clearance between valves and pistons, 6,600 to 7,000 RPM is the limit for track racing; 7,600 for drag racing. Balanced, blueprinted and tricked-up big-block Chevys have been taken to RPM's in excess of ten grand, but this is a short sure way to saw one in half. Even the handful of professionals who occasionally experiment at such wondrous "R's" are quick to admit that they occasionally lose an engine. For this reason, and because no one sells engine insurance, pay the "premium" required for a rev-limiter system.

There are two basic types of RPM limiters: electronic and mechanical. Mechanical types operate from a cable connected to the distributor tach drive. If a mechanical tachometer is also used, a dual-drive connector is required.

Stahl Associates offers Jones mechanical tachs and a mechanical rev limiter/ignition shut off. The coil primary wire is wired to the limiter switch. When the pre-set RPM is reached, the ignition is turned off until the engine slows to 100 RPM below the shut-off point. You should be aware that the mechanical type shut-off shuts off the ignition completely until the RPM get back to the desired level, which may not be what you really want . . . especially on a super-charged engine, because the manifold can get well loaded with fuel, as can the exhaust system, during the time that the ignition is completely turned off. This can be perfectly o.k. on a drag-race machine with open headers, but an electronic type limiter will probably be better for a street/strip machine that uses mufflers. Why? Because the drop-dead mechanical shut off loads the mufflers with fuel and it is not unusual for the mufflers to be completely blown off of the exhaust system. How are you going to explain that to the passing gendarme?

In the area of electronic rev limiters, ARE, Inc. probably supplies more units than all of the others. Their Revgard RPM Limiter and a similar unit with a built-in "dwell stretcher" can be pre-set to any desired RPM. When the engine reaches that RPM, some of the ignition pulses are removed to hold the engine at the RPM limit. Two wires connect the ARE Revgard unit into the ignition system.

Jones Motrola tachometer is cable-driven. Tell-tale is optional.

Turbosupercharging
free HP from your exhaust

Two Rajay turbosuperchargers with a single Q-Jet pump 425 HP at 3,800 RPM from the Daytona Marine 427 CID Chevrolet Turbo-Jet engine. It has 7.5 c.r. and cranks out 460 lbs. ft. torque at 2,800 RPM; 590 lbs. ft. at 3,800 RPM. The engines are winners in offshore racing.

Keep your eye on this low-cost technique for adding HP. With the current fight to control emissions, the turbosupercharger may end up as one of the few ways you'll be allowed to add HP to your street-driven automobile. The accompanying graph shows what Chevrolet found when testing a 482-cubic-inch marine engine—one of 25—made for Kiekhaefer. Note that the power is available 'way down. Revs were not needed to get the maximum power.

Turbosupercharging has become an extremely important factor in fast engines since the mid-70s. HPBooks has published a book on the subject entitled *Turbochargers*. Author Hugh MacInnes, Chief Engineer for Rajay Corporation of Long Beach, California has designed more types of turbosuperchargers than any other engineer. He has put his expertise on paper, with descriptive drawings and photos of installations on engines ranging from motorcycles to big-block Chevrolets. He tells you how to select turbocharger size, ways to install it, and controls for making sure that you get boost when you want it.

If you'd like to discover how two turbosuperchargers can turn your big-block Chevy into a snorting "moose motor," get a copy of *Turbochargers*. See the inside back cover of this book for ordering information.

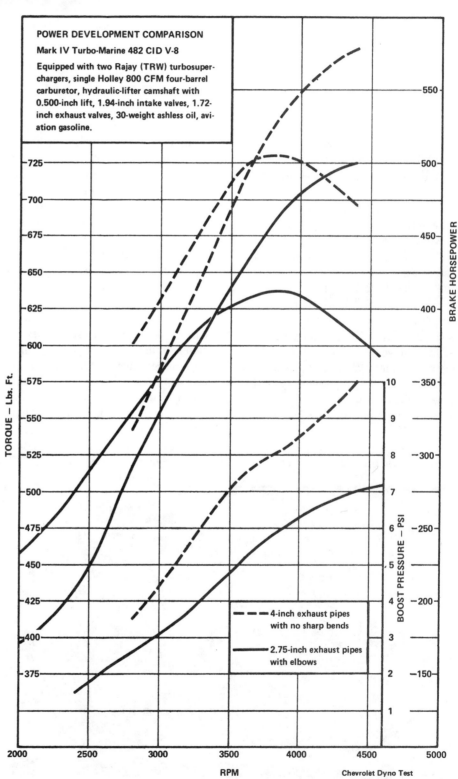

POWER DEVELOPMENT COMPARISON
Mark IV Turbo-Marine 482 CID V-8
Equipped with two Rajay (TRW) turbosuperchargers, single Holley 800 CFM four-barrel carburetor, hydraulic-lifter camshaft with 0.500-inch lift, 1.94-inch intake valves, 1.72-inch exhaust valves, 30-weight ashless oil, aviation gasoline.

--- 4-inch exhaust pipes with no sharp bends
— 2.75-inch exhaust pipes with elbows

Chevrolet Dyno Test

Clutch & Flywheel
these work best

We won't go into the science of flywheel and clutch selection—not even into the modifications of transmissions and rear ends. All this is the subject of constant experimentation on the part of drag racers and the makers of clutches and flywheels. We can tell you what Chevrolet has done for you and has available in their parts bins.

Big-block Chevy engines produce so much torque that it is difficult to make a single-disc clutch that will hold it without making the spring pressure so stiff that it would take two well-developed feet to operate it. When the engines are hotrodded so that the power output is increased, the average single-disc clutch assembly starts slipping after only a few hard runs. Slipping clutches are no real problem to the professional pro-stock racers because they have been known to set up their small-diameter clutches to give a small amount of slip on engagement so as to cushion the drive train from the shock which it would otherwise receive. However, such professional racers are hardly worried about the costs of rebuilding the clutch and carefully examining the transmission after every meet. You probably are concerned and would like to avoid rebuilding things any more than necessary. After all, it takes time to haul that big lump of an engine in and out of your chassis—and parts—even stock ones—aren't all that cheap. If you are building on a budget, then the best way is to put the right parts in to start, rather than just hoping that everything will work out o.k.

If you have a Turbo Hydra-Matic 400, you've been laughing as you read the previous paragraph because you have solved your clutch problems. And, if your converter has the correct stall speed, you are probably going as fast or faster in the quarter mile than you did with the clutch-and-four-speed arrangement... with no further worries about the clutch. You've wrapped your automatic in a protective shield blanket, of course.

But, if you are living with a transmission and clutch, you may want to consider a few changes when you take things apart to install your Ansen or Lakewood steel scatter shield. The accompanying table shows you a little about what goes on in the clutch department and should help you to understand what Chevrolet offers. You'll note that some 454 Corvette engines are equipped with a dual-disc clutch. This clutch and flywheel arrangement can be added to any 454, or by buying the non-counterweighted flywheel (all of this is listed in the HD Parts List), you can install one of these dual-disc units in any 366, 396, 402 or 427. If you already have a 14-inch-diameter flywheel, then the dual-disc parts will slip right into position. However, if your car has the smaller clutch and flywheel, you will have to buy a new bell housing, 3899621. You'll also need a new starter housing, 1969309.

The dual-disc clutch is not recommended for drag racing where fast shifts are essential. The extra inertia of these units makes the synchros work harder. However, for long life on the street where super-fast shifts are not all that essential, the dual disc has a couple of good points in its favor. First, the milder pedal pressure makes life a lot easier for anyone who drives the car. You won't have to set up a special leg-building program at the local gym to be able to operate the clutch pedal. And, your wife or girl friend will be able to drive the car without complaints—about the clutch, in any event. Secondly, the clutch will probably never wear out. As long as you keep driving Chevys and the flywheel pattern does not change, you can keep swapping the clutch from car to car.

Buy one and use it on your Chevy cars from now on. You could probably will it to your grandchildren and it would still not wear out a hundred years from now. Chevy's dual-disc clutch literally ends all worries about clutch replacements. The first cost may be high, but it's worth it!

If the lofty price of the dual-disc clutch does not appeal to you, you can get the 10.5" L-88/ZL-1 single-disc clutch. This will hold onto most 396, 402 and 427 engines fairly well and its reasonable pressure will allow you to operate it without continual leg cramps. Use this clutch with the 153-tooth 12.75-inch 15-lb. flywheel 3991406 (was 3866735). To use this (or any other 153-tooth flywheel) requires using 1968122 starter housing and 3858403 bellhousing. Both 10- and 26-spline discs are available.

If you want to install a 168-tooth 14-inch-diameter flywheel, the required starter and bell housings are numbered 1969309 and 3899621, respectively.

In general, the Chevrolet clutches offer better quality at lower prices than the hotrod clutches. Here again, don't be misled by the decals on your hero's car. The clutch that he's running may be a far different animal than the one that you can buy as a standard part.

Should you decide to experiment with super-stiff non-Chevy single-disc clutches, you may end up having to buy a hydraulic clutch-actuating mechanism to be able to drive the car comfortably. In this instance, look at the parts in the bins for the Chevrolet trucks.

When you add up all of the factors, the total real cost could cause you to rethink the problem. A Turbo Hydra-matic 400 or a dual-disc clutch could be your answer. Incidentally, if you go to the wrecking yards for a Turbo Hydra-matic, be aware that asking for Chevrolet "anything" automatically jacks up the price. The kind of transmission that you should be looking for has the vacuum modulator on the right side toward the bellhousing end of the case. This says that the unit is a Turbo 400. You do not want a Turbo Hydra-matic 350 with the vacuum modulator pointing backward at the U-joint end of the case. Any GM Turbo Hydra-matic 400 will work—Buick, Oldsmobile or Pontiac. Adapter plates are available to mount BOP transmissions to Chevy engines. If you find a broken-case BOP 400 at the right price, consider buying it and installing the guts in a new Chevy case.

For more on clutches, flywheels and related components get HPBooks' *Clutch and Flywheel Handbook*.

Three stock flywheels (there are many more). Left is 14-inch-diameter 168-tooth for use with the dual-disc pressure plate and 10-inch clutch discs. Center is 396/402/427 12.75-inch-diameter, 153-tooth flywheel. Right is 454 12.75-inch-diameter, 153-tooth flywheel with large counterbalance on backside (arrow).

CLUTCH SPECIFICATION TABLE

Year	Car	Total Spring Pressure (pounds)	Disc Effective Area Size (inches)	(square inches)
1966	325/350 HP Corvette	2100-2300	10.0x6.50	90.7
1966	390/425 HP Corvette and all passenger cars with big blocks, except L-72 which had 2600-2800 lbs	2300-2600	10.5x6.50	103.5
1971	365 HP Corvette 402 CID Camaros, also 1970 390/400 HP Corvette	2450-2750	11.0x6.50	123.7
1971	425 HP Monte Carlo and 1970 435 HP Corvette	2600-2800	11.0x6.50	123.7
1971	425 HP Corvette (dual-disc)	1600-1800	10.0x6.00	201.1
1967	L-88 Corvette	2900-3100	10.5x6.50	103.5

Bore/stroke info
unusual facts you may have overlooked

Which bore and stroke combination should you use in your next engine? That's the subject of more bench-racing sessions than can be counted. We've participated in a lot of them ourselves.

The larger the engine—the more horsepower—usually. But, the HP may not be "made" exactly where it's wanted in the RPM range. So, bore/stroke combinations are often jockeyed around and nearly always at great expense. For street use, boats and drag racing, the chances are good that you should use the longest stock stroke and biggest stock bore which you can afford within the racing rules and the size of your bank account. That's just what your competitors will be doing; count on that! Cubic inches are always the cheapest HP adder and the most trouble free.

A displacement table on page 100 describes some of the bore/stroke combinations which you may want to consider. It does not include the 4.44-inch-bore Can Am block, which gives 430/465/496 CID displacements, depending on which crank is used. In general, the shorter stroke tends to move the HP upwards in the RPM range and a longer stroke tends to produce the HP at lower engine speeds.

The accompanying graph shows test data obtained in a 1968 comparison at Chevrolet Engineering—part of a program to evaluate pieces for a 430 CID engine which was to have replaced the 427. Although the shorter stroke gave slightly more power and even a bit more torque above 3,000 RPM, it was a "dirty" engine from an emissions standpoint. So, the longer stroke 454 became the "Mr. Clean" replacement and we were all treated to the availability of a stock 4-inch-stroke crank from the dealers' parts bins.

The 3.47-inch-stroke cranks did get used—with special long rods—in the 4.44-inch-bore Can Am blocks used by most of the teams in 1969. HP was about 650. By 1970, the engine builders were stuffing 3.76-inch cranks into the blocks to get 465 CID and 670 HP. A few started using the 4-inch 454 cranks. Then came 1971 when the 454 crank became almost mandatory in Can Am engines, giving 495 CID and 710 HP. In this no-holds-barred racing, bigger is not only better—it's essential to keep your competition in sight.

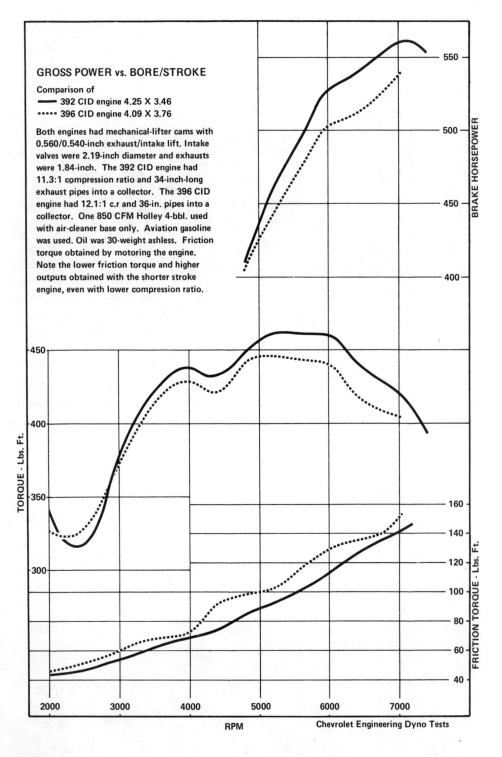

GROSS POWER vs. BORE/STROKE
Comparison of
— 392 CID engine 4.25 X 3.46
····· 396 CID engine 4.09 X 3.76

Both engines had mechanical-lifter cams with 0.560/0.540-inch exhaust/intake lift. Intake valves were 2.19-inch diameter and exhausts were 1.84-inch. The 392 CID engine had 11.3:1 compression ratio and 34-inch-long exhaust pipes into a collector. The 396 CID engine had 12.1:1 c.r and 36-in. pipes into a collector. One 850 CFM Holley 4-bbl. used with air-cleaner base only. Aviation gasoline was used. Oil was 30-weight ashless. Friction torque obtained by motoring the engine. Note the lower friction torque and higher outputs obtained with the shorter stroke engine, even with lower compression ratio.

Chevrolet Engineering Dyno Tests

Blueprinting & assembly tips
time, care & patience get the job done right

Precision bore gages are used by all professional engine builders because they are faster to use and aid in making accurate measurements. Here Colin Beanland checks the bore on a Reynolds Aluminum all-alloy block as used by the McLaren Racing Team. These blocks can be used to build really big engines—up to 524 cubic inches—as discussed in the text. Thomas Bedford photo courtesy Reynolds Metals.

Getting it all together

Now that we have the building blocks for the various big-block engines fully described, we'll get into how it all goes together. Any of the following detail is meant to apply to any displacement big-block Chevy unless otherwise noted.

Putting a big-block Chevy engine together in a winning combination is TEDIOUS! No other word describes it. If you are in a hurry — bury or burn this book — because it will have been of no use to you. It is amazing how a man will put hundreds of dollars into open-chamber heads, special cam, double-pumper carb, forged pistons, new rods — and so forth — and then insist on hurrying through the marriage of these expensive parts. The unfortunate ending to this story is that this same man is quickest to blame the Chevy factory or speed-equipment makers for turning out such "lousy" merchandise. The old adage that haste makes waste can be no better applied than to the horrendous task of assembling a high-performance engine. It ain't easy! One of the best, shortest and most meaningful comments on assembling parts to make a really good engine comes from Hal Klieves of Race/Chek in Pompano Beach, Florida.

"The best procedure for building any engine is patience, care and constant rechecking."

Engrave that statement on your headbone. Rechecking is the one outstanding trademark of the professional racing-engine builder. The average mechanic is quite content to bolt an engine together *once*, but his patience turns to anger if he ever has to take parts off or go back to do the job or some part of it over again.

Extra assembly – *even three or four times if that's what it takes* – measuring, disassembly, modifying, reassembly, rechecking and so on can be satisfying or frustrating, depending on your temperament. If you are easily frustrated and become angry when setbacks thwart your progress, forget about laying a wrench on the big-block or any other engine. Pay the cost and have the work done by an expert because engines jammed together hurriedly will never win races — or reliability contests, either.

This section is possibly misnamed because it also covers a great deal of information for preparing for the job at hand so that the frustrations will be reduced when you are doing the actual work. Additional details are provided on torque measurements and clearances so that you'll have these handy. Careful attention to details ahead of the time when you pull the engine can definitely shorten the time required to get the engine back together and into your automobile again. Such items are important whether the engine is used for competition — where it is important to get ready for the next race — or for transportation where the car must not be kept out of service for long periods. Many enthusiasts have difficulty getting all of the parts together at the right time so that the work can proceed without delays once it is started. You should make your shopping list carefully according to the engine which you are planning to build. Order all of the components which will take time for delivery. Note that high-performance, heavy-duty and off-road parts in the heavy-duty parts list may not be instantly available from the Chevy dealer's parts bins. Wait for the special parts to arrive and do whatever work is required on these parts — before even taking the engine out of the chassis. Pistons, rings, bearings, camshaft, crankshaft, valve springs and retainers — and many other parts — often take some time to arrive after you have ordered them.

Blueprinting

Despite all of the long-standing mystery and confusion associated with the term, blueprinting simply means rebuilding an engine to an exacting set of specifications. Some of these details are those originally set down by the factory; more often than not, however, the tolerances involved in rebuilding any high-performance engine vary slightly from factory-rebuilt specs. The thoroughness and care with which each component (and this goes for the last bolt you tighten) is measured and inspected pays off not only in added horsepower, but also in extended engine life.

Can an engine be blueprinted in a small garage or hobby shop? Yes, but! Unless you are well-equipped with tools and some special equipment, the job will be extremely difficult and very slow for the beginning engine builder. Because of special equipment needed for some steps, some operations may have to be by-passed, but for the most part an engine can be blueprinted in most any garage or workshop maintained by a performance enthusiast who is serious about his hobby.

Miking the deck-to-main-bearing-saddle distance on an aluminum ZL-1 block. The bolt-on plate seals rear cam bearing in aluminum big blocks.

An align-boring machine is good for only one job—and that's align boring—period! So, you'll not find one in every automotive machine shop. In fact, the majority of big-block Chevys are quite "straight" as they come from the maker. The only exception is when installing aftermarket steel main caps. Nevertheless, don't align-bore unless the job is required. Save your money!

If all of the measurements check out when the block, crank, etc. are returned from the machine shop, fine. If not, then the blueprint job is in deep water and there may be no way to save it short of buying new parts and starting over again.

Bear in mind that any steps left out, any variations in the clearances and specs means less horsepower and shorter life for the engine. Just before you decide to cut one little corner here or there, picture your engine with a rod sticking out the side of the block or a

130

valve bent by a piston. Then consider what that will do to your bank account! The added time and care applied to an engine now can mean thousands of extra miles later on.

Cleaning the Block

The first step in blueprinting requires disassembling the engine down to the last bolt, even if you are starting with a brand-new engine. Whether the engine is new or used, remove all of the soft plugs and the cam bearings. New cam bearings can be reused if you take them out carefully with the correct driver—which you should buy for your tool kit anyway.

All of the oil plugs must be pulled from the side oil gallery. Check these to make sure that all of the holes are drilled and finished. Sometimes there will be a lot of sealing material left behind these plugs and this must be removed.

If you are tearing down an engine that has been assembled with Teflon tape, make sure that you capture every last bit of this stuff when removing the plugs. A small pick or wire can be used to remove pieces of tape from the threads. Go through all threaded holes with the correct size tap to clean up the threads.

While you are cleaning the block, don't forget to clean the heads. Have the block and heads "boiled out" or "hot-tanked." Aluminum heads and blocks take special care and must be cleaned in a *non-caustic* solution. Most auto-parts houses, automotive dealerships or automotive machine shops can do this job—or tell you where to get it done.

A small bit of strategy can be used here, so work it into your plan of attack. Take the block to be boiled out on a Friday afternoon if the place of business involved is closed on Saturday. Check with the man in charge of the operation and request that your block be left in the cleaning solution over the weekend. Or, make arrangements for the block to "cook" for several days before it's pulled out.

If you don't build engines on a regular basis (at least two a year) another sly move can be effected by asking the man who boiled out the block if he'd mind taking a look at the block to see if he can find any cracks or problems which could plague you later on. This is done on the theory that two sets of eyes are better than one, coupled with the fact that a man who works at this all day can often spot items in an engine that a newcomer to the field wouldn't see or even understand. Look for fine hairline cracks in any of the main webbing, and double check all of the bolt holes for pulled or stripped threads. The chances of finding any cracks in the main-journal webbing are remote, but look now before you cart the block home. Needless to say, if you find *any* cracks, pay the boil-out bill and point the block toward the nearest trash barrel.

Magnaflux and Zyglo

Professional engine builders will always Magnaflux the block and caps to make sure that they are not overlooking cracks which could turn their expensive work into a pile of junk — perhaps in the heat of competition. If you get to Magnaflux equipment, this is cheap insurance against making a big investment in a block that should be broken up and melted down again.

If yours is an aluminum block, the main caps should be Magnafluxed and the rest of the block should be Zyglo-ed.

Align-boring

The next check which you should make requires a machined-steel mandrel approximately 27 inches long with

On iron-block 396/427 Chevys an oil passage runs alongside the skirt. Remove all plugs leading to this passage before cleaning the block to ensure flushing all residue from this passage.

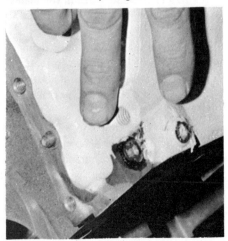

2.9365-inch diameter (0.0005-inch undersize for the minimum-diameter main-bearing saddle bores). Lay the mandrel into the bare block—with no inserts in the bearing saddles or caps. Torque the caps into position. If the mandrel can be easily turned by hand, the block is straight enough for all practical purposes. In fact, even if the mandrel turns slightly snug in the bearing bores, it is probably best not to align-bore the block because your chances of getting a good align-bore job from the average automotive machine shop are next to impossible—even if the owner does it himself. We are not referring to the machine shops that specialize in racing-engine construction, of course. But if you have ever thrown a block in the junk pile because you could not get the crank to turn in its bearings after the local automotive machine shop had done an "align-bore" job—you'll know what we are talking about. We've seen it happen—and the job, even though it wrecked the block and caused the owner a lot of grief—was never free. Don't think that the friendly small-claims court will stand behind you. After all, the local businessman's word must be truth, right? So unless you live where there is a good access to top machinists who speciaize in race-engine preparation, avoid getting an align-bore job. Many reputable engine builders report that the big-block Chevys have sufficient beef that they rarely need align-boring.

Even if a machinist has the Quik-Way machine which locates the block according to the crank centerline, you are still not sure of what quality work you will get because it is the operator—*not the machine*—who makes or breaks the job.

Some people think of align-boring as a rather simple job that any automotive machine shop can do. *Not so.* Align boring is done to ensure that all main-bearing saddles lie on a common axis. To accomplish the operation, a slight cut is taken from the mating surface of each main-bearing cap. The caps are bolted to the block and torqued in place. A boring bar is then run the length of the block through all of the main-bearing bores. There are machines built especially for the purpose, but not all are alike. There are a lot of so-called align-boring machines which are still being used, even

Machined straight edge can be used in a preliminary straightness check of the main-bearing saddles. Here an aluminum big block is being checked.

though the holes that they make and the method of locating the equipment to the block, leave a lot to be desired.

If you have a block align-bored, remember that the cam-to-crank centerline distance may be reduced—with consequent timing variations due to camshaft "slop" unless the machinist only takes a thousandth or so out of the block in the boring pocess. If you are align boring a block in which the main bearing insert has "spun," and get the block bored with a 0.005-inch cut in the block side, you can get a Cloyes True-Roller chain/sprocket set designed for the closer center distance. These sets must be special ordered. When you add up all of the costs which are involved, you may find that it is cheaper to junk the old block and start with a fresh new one.

Align boring simply cannot be accomplished in some small towns—and in many big ones, too. A local automotive machinist or garage should be able to tell you where the work can be accomplished, but be extremely wary. Prices vary from shop to shop, but expect to pay about $125 for align boring.

Incidentally, if you don't have a mandrel, a straight crankshaft will give a good indication of how straight the block is. If, when you lay the crank in the bearing inserts and tighten the caps — with no seal — the crank turns freely by hand, the chances are good that this is going to be an o.k. block for a street machine. Details for checking the crankshaft for straightness are provided later in this chapter.

Decking

On all-out racing engines, block preparation and subsequent machining can be quite expensive because of the large amounts of time which are inescapably involved. Although this detailing is not necessarily required for a strong-running, street-only, large-block Chevy, the true racer would never think of overlooking it. That is the "decking" of the block. This operation gets the distance between the deck and the crank centerline equal at each end for each cylinder bank. It also makes the decks 90° to the cylinder bores and establishes them at a height above the crank which gives the desired deck clearance between the piston flat surface and the cylinder head.

Big blocks are usually out by as much as 0.008 inch from one end to the other and have been known to stray from parallelism with the crank by as much as 0.027 inch. Measurements can be made with a long depth mike from the deck to a mandrel installed in the main-bearing saddles, or with a large-throat mike from the deck to the bearing saddles at each end. Be sure to record the figure that you end up with on the corrected deck height so that you could build another block just like it if you had to make a replacement.

There are a couple of ways that you can go about the next sequence of events to get the deck height established and the decks parallel to the crankshaft centerline. If you are planning to reassemble the engine with the same parts, *check the deck heights before you take the engine apart!*

Clean the carbon off of the flat portions of the piston tops. Use a dial indicator in a bridge-type holder or a depth mike with parallels (a couple of ground tool bits work just fine). Measure the deck height of each piston and record these dimensions so that you can refer back to them later on. Note whether the piston is above or below the block deck. If a piston is above the deck, put a + sign in front of that number. Mark the piston and rod assemblies before you take the engine apart, preferably stamping them with numbering stamps. That way you'll know "where you're at" if one of your buddies cleans the parts when your back is turned.

After the engine has been disassembled and the block hot-tanked as already discussed, check the main-bearing saddles for freedom from cracks and for true alignment as previously described. If line boring is required, do it now. If you are going to use the same pistons and rods, get the block decked to provide the desired relationship of the pistons' flat surfaces to the block deck.

But, you say, all of the pistons did not have the same deck height. Now you are getting down to the real nitty-gritty. You must use the lowest piston as your reference. Deck to *that* piston. The other pistons' flat surfaces will have to be machined to lower their decks to equal the lowest of the bunch. This can also mean that the piston domes will have to be machined off a bit, too, to keep the compression equal in all cylinders. Measuring piston-dome volume and calculation of compression ratio is handled in the cylinder-head chapter.

Deck clearance should be from zero (piston flat even with block deck at TDC) to 0.005 inch with the piston centered in the bore. This assumes that you are using the 0.040-inch-thick Victorcore gasket as supplied by Chevy. Simple arithmetic tells us that this provides a 0.040-inch clearance between the flat on the piston and the flat portion of the combustion chamber — if there is zero clearance — that is, piston flat even with the block deck. Or, you'd have a 0.035-inch deck clearance with the piston protruding 0.005 inch from the block deck.

If you should use a thinner gasket, then the piston will have to be lower than the block deck to provide a minimum of 0.035 inch deck clearance when the head is installed.

Keep in mind the things which affect deck clearance. These include compression height (pin center to flat part of piston top), rod length (center of big end to center of small end), position of crank in the block (it will be raised slightly if the block is line bored), and crankshaft stroke at each rod journal.

If you are going to bore the block with a machine which references off of the deck surface, the deck must be trued parallel with the crank centerline prior to boring.

If the block deck is crooked—then the bores will be off. Although the factory uses automatic equipment which lines up on crank centerline, the bores are not always straight and true with the crank centerline. You can get the bores closest to straight by lining the boring machine up in the center of each cylinder, half way down the bore. This provides an average location which will be more nearly correct than if the boring machine is lined up with the top or bottom of the bore.

If the block is being bored with a machine which references from the crank centerline without regard to deck position, then you can skip this preliminary parallel-decking operation.

Once the block is bored, assemble the engine loosely with the parts which are to be used and check the deck heights at each piston. Note the measurement on each piston top with a marking pen. If you merely had a preliminary cut made to parallel the decks to the crankshaft, a further cut may be required to get the deck correct for the piston which sits lowest in the block when it is at TDC. Once the deck is correct for the lowest piston, check the rod lengths and compression heights to see whether you can find a different combination which will reduce the amount of machining required on each piston. Because piston machining will nearly always be required in getting the deck height correctly established, you can see that any piston balance operations should be delayed until after deck height has been established. And, piston notching may also be required, as described in the camshaft chapter. You may want to make the piston-to-valve check for clearance at this time so that all machining operations on the pistons can be done at the same time.

Correcting the deck is best done in a block machine which grinds the deck surface. A large milling machine can also be used. Incidentally, a dead-smooth surface is neither wanted nor needed. The surface should be rough enough so that you can catch your fingernail in it.

More block work

With a handful of grinding wheels and a 1/4- or 3/8-inch drill motor or high-speed grinder, go to work eliminating the rough spots all over the cylinder-block interior: the tappet chamber beneath the intake manifold, the crankcase and the timing-gear cavity. Most engine builders start grinding by using various shaped stones—then go back over all of their work with some of the fine-grit paper sanding sleeves. This gets rid of any small bits of sand or casting which might jar loose, work their way into the oil system and "do in" the engine. We don't want the oil to "stick" to the rough surface of the stock tappet chamber and crankcase either. Polishing and then painting (later) aids oil drainback to the sump. With this polishing and deburring the block is now ready to be fitted with the pistons. All pistons in a set will vary slightly in size and for this reason should be fitted individually into their respective bores per the piston clearance specifications in that chapter.

Two views of the four-bolt main caps showing the two types of bolts which are used. Center bolts are double-ended to mount a windage tray/baffle. These require a deep socket or an offset adapter to allow tightening with a torque wrench to the recommended 110 lbs. ft. Small arrows cast near the center of the caps should always point forward. Main caps must always be reinstalled in their original positions on the block.

A stock cross-hatch pattern is desirable on any hone job, no matter what grit stone is used. Generally, the coarser the stone, the harder the ring—except in the case of moly rings. A 300-grit hone is usually used when running a cast-iron ring, and a 180- or 200-grit hone is standard for chrome rings. However, a 500-grit hone is used for moly rings because any hone coarser than this leaves a surface finish which tends to chip the moly out of the ring.

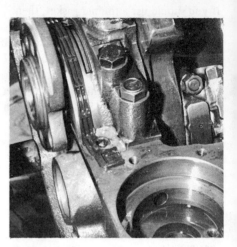

Pro Stock engine builder Wally Booth uses Silastic RTV sealant where pan gasket joins rubber seal. A tube of this stuff is like money—it can be used in a lot of places. Sears sells it as Silicone Adhesive Sealant in clear, metallic, black and white. Regardless of color, it's all the same good stuff.

Assuming the engine block is on a stand of some type, move it to an area which you don't mind being doused with soap and water. Mix up a dishpan or bucket full or solvent (from the local service station), dishwashing detergent (Tide, Cheer, etc.), and hot water. Start scrubbing that block with stiff bristle brushes and the solution. Give it the old "Dutch rub" inside and out. Industrial supply firms and some automotive and speed shops sell small bristle brushes attached to very long wire handles which will allow cleaning of the oil galleries over their full length. If you can't find such brushes, a local gun shop will fix you up because the gun-barrel-cleaning brushes work just fine in an engine. After you have the engine (and yourself) thoroughly soaked, sudsed, and scrubbed, apply the nozzle and hose to get rid of all the suds and, hopefully, any small particles which might still be around. You may prefer to haul the block to your local do-it-yourself car wash to use the high-pressure spray equipment there. This approach works well. While the block is drying take another close look all over for hairline cracks, a bit of sludge hidden in an oil passage leading to a mainbearing—anything which needs to be taken care of now. Don't hesitate to use a magnifying glass—especially if you're over 30! Now that you've gone this far with the project, mix up a new batch of detergent—without the solvent—do the whole job a second time. Rinse away the suds and let the block stand in the sun or under a heat lamp or two to aid in drying. If an air hose is handy, use air pressure to aid the drying-out process. From here on in, you are working with a clean block and you must take every precaution possible to keep it that way. This includes speeding the drying process and getting the block under a large towel or plastic covering to keep it dust-free until you are ready to paint it.

Bob Joehnck points out that block painting certainly ensures that no loose sand will get into the engine, but he claims that it is almost impossible to paint blocks commercially—because it is too time-consuming. Here is another place where the occasional engine builder's "free labor" can ensure a good job.

Actually, if the costs of sonic cleaners were not so fantastic that only large industrial firms and aircraft or missile builders can afford them, the sonic cleaner would be the ideal way to clean the block. Then there would be no need for painting. Some cleaning tanks (not sonic ones) have jet nozzles that direct strong streams of cleaning solution against the various parts of the block. If such are used with lots and lots of elbow grease and subsequent cleaning in household detergent and hot water, Joehnck feels that this is adequate.

Typical brush needed to scrub the oil gallery (arrow).

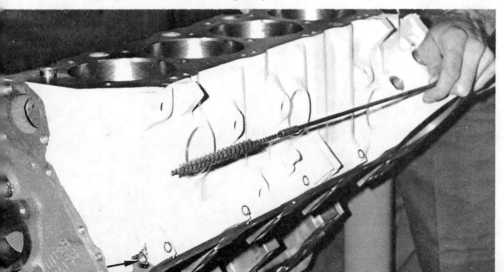

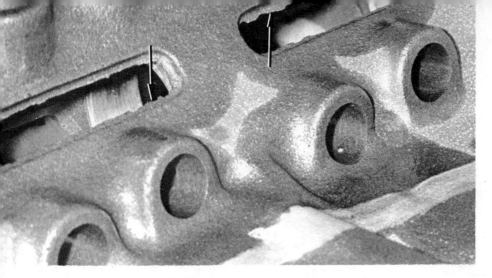

Arrows indicate casting "flash" which should be ground away.

Glyptal by General Electric is the trick for painting the inside of a block. Electrical motor rebuilders use it—so shop there.

All bolt holes should be chamfered in a high-performance engine to eliminate pulling threads when the torque load goes up.

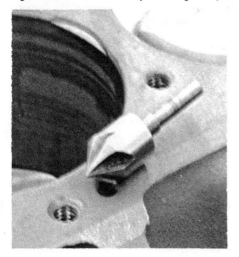

Painting the inside of a block serves two functions. The paint's smooth surface speeds oil return to the sump. And, the paint goes a long way toward holding any grit in place which might work loose from the casting even after your polishing and scrubbing operations. You'll need a small brush—one one-inch wide will be fine—and a pint of Rustoleum or Glyptal. Rustoleum can be purchased at most any hardware, paint store or lumber yard which sells paint. Glyptal may be more difficult to find but is well worth the effort. It is a material manufactured by General Electric for coating electric windings in generators and electric motors. Thus, your search for this should begin in the Yellow Pages for a firm specializing in rebuilding large electrical motors. Before dipping into either of these coatings, move the block into a dust-free area and rig a sun lamp or two over the block to aid the flow and to set up the coating harder than can be achieved by drying at room temperature. Paint the area in the front of the block which will covered by the timing chain cover. Paint the entire tappet chamber which is to be covered by the intake manifold. Flip the block over on the engine stand and cover areas between the cylinder bores. At the back of the block, paint the area which will be covered by the bell-housing. This painting is a rather slow process—or should be—because you should not brush any paint on the machined surfaces, such as the cylinder bores, head-mating surface or any of the machined surfaces adjacent to the main-bearing journals. Let the paint harden for at least 24 hours before getting back to the block.

If at all possible, leave a couple of sun lamps pointed at the oil gallery and the crankcase to harden the paint.

If you are spray-painting your block with Rustoleum, use this procedure to get the best job: first, fog on a light coat and let this set for 24 hours; next, spray on a slightly heavier coat and let that set for 24 hours. Finally, spray on a heavy coat and let that set for 24 hours before removing your masking tape. Incidentally, corks or rubber stoppers are very helpful in keeping paint out of the lifter bores, distributor hole, etc. Don't hesitate to use them if

Chamber beneath intake manifold should be painted to aid oil return to the pan. Notice that paint is carefully brushed around but not in the lifter bores.

Slim Carter of Lake Speed in Mentor, Ohio displays the steel plate he torques atop big blocks during honing operations to simulate loads applied by head bolts. This allows hone to work on the block as it will be distorted when it is running.

they are available, as they speed things up as compared to masking tape.

There is only one valid reason for painting an aluminum block Chevy—to create the impression that "my motor is just a reglar 'ol motor." At times this can come in handy; if you're not apt to use such tactics—don't bother. The aluminum block is of very fine quality and there is almost no chance anything will flake out of the casting and contaminate the oil. The casting is so smooth that little if any gain in oil drainback could be expected. However, if you are bound and determined to paint the block, clean it, wipe off the area to be painted with Prep-Sol or other surface-preparation cleaner and then prime the surface with zinc-chromate primer. *This is a must!* After this has dried, the Rustoleum or Glyptal can be applied as with the iron block. You might want to paint only the exterior of the block to give the appearance of a "bucks-down" racer.

Block Plugs

Some mechanics tap the block "freeze-plug" holes with a 1-1/2-inch pipe thread and install pipe plugs from an electrical supply house. These are installed with Loctite to eliminate any further worries about lost or leaky plugs. Others swear by the Neoprene expansion-type plugs which tighten in place with a nut. Still others use the stock cup-type plugs epoxied or Loctited in place. Aluminum plugs are threaded into the aluminum blocks and these are usually welded in place before the engine leaves the factory. If you get an alloy block which has not had the plugs welded, a quick bit of work with the heliarc "wrench" will permanently leakproof the area.

4-Bolt-Main Conversion Kits

Should you have a two-bolt-main block and be wondering what to do with it, the chances are good that you should run it as is for a street machine. If you are going to get into serious competition, then it is possible to use the block with a four-bolt-main conversion kit. These kits require align boring after the block has been drilled and tapped to mount the new caps. However, consider the total costs because a new high-performance block including new cam bearings for any standard displacement costs approximately $550. This could be far less than the cost for a conversion kit, align boring and a new set of cam bearings, plus a bore job if that's needed.

When you buy a fitted block, the deal gets even better. That arrangement gives you pistons, rings, pins and both main and cam bearings. Some of the factory-direct deals are so good that we wonder why anyone ever considers buying anything other than stock parts.

Balancing

Lots of engines are assembled and run with stock parts without being rebalanced. This is especially true where the assembled engine or short block came straight from Chevrolet. Isn't that a surprise? Balancing is one more of those oversold speed-merchant services which separate you and your hard-earned $$$. However, if severely flycut pistons are used or the rod weight is changed, the engine should be rebalanced.

As part of finding more about balancing, we checked with Chevy Engineering to find out whether the old racers' tale of truing up and rebalancing the harmonic balancer/damper made any sense. We found that these are balanced as they come from the factory, even though they may be slightly out-of-round at the OD. If you "true them up" to be perfectly round, you destroy the factory balance job and you do have to rebalance the part. If you were wondering how the stock ones last without any special treatment, consider that a million hours of durability testing on engines with stock balancers showed no balancer failures. So, here's another place you can save by not trying to out-engineer what has already been done for you with cheap factory parts.

If you are having the engine balanced, get the harmonic balancer and clutch/flywheel assembly balanced at the same time — preferably with these units attached to the crankshaft as they will be run in the completed installation. This is essential with the 454, of course, unless the crank has been "internally balanced" by adding Mallory metal to the counterweights.

Caution must be used in rebalancing a harmonic balancer. Excessively deep balance holes will weaken the inertia ring and can cause failure at high RPM. Chevrolet recommends a maximum balance-hole depth of 1/4-inch using a 1/2-inch-diameter drill on the front and/or rear of the rim. These should be on a 3.62-inch radius on an 8-inch-diameter balancer. Most important, the balancer rim should be tested to prove that its hardness is 187-241 Bhn. If it does not meet the hardness specification, get a new balancer

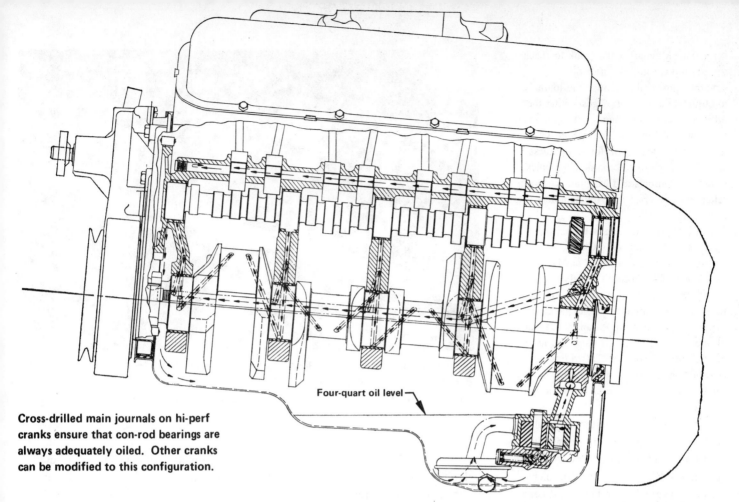

Cross-drilled main journals on hi-perf cranks ensure that con-rod bearings are always adequately oiled. Other cranks can be modified to this configuration.

Crankshafts: Not so groovy!

Preparation of the crankshaft is another area where the guy building engines at home gets left out in the dark when it comes to an all-out racing engine. In this case, the crankshaft should be Magnafluxed. Some would try to tell you that the crankshaft should be "fully counterweighted" by adding extra weights at the center of the crankshaft, but the Chevy engineers did not think it was required or they'd have built the crank that way to start with. This is another area where you can examine Can Am and top drag-racing engines and discover that wise builders use the cranks as they come from Chevrolet.

Magnafluxing is a dye-penetrant electromagnetic process for detecting flaws or cracks in metal. Few machine shops offer the service—but if the service is offered in your area—take advantage of it. The cost involved, as opposed to the cost of replacing an entire high-performance engine, is a pittance. We mention this because we can think of little to save in a big-block Chevy once the crank "lets go."

All oil holes in main and rod bearing journals should be slightly chamfered and deburred with a high-speed hand grinder. First use a fine grinding stone, then switch to an emery-paper cone and finally polish it with a rubber-impregnated Cratex tool in the grinder. However, to observe Cratex's RPM limit, you may have to switch to a slower-RPM electric drill for the final polishing step. Lock that crank down tight and hold the grinder before you start this operation or you could badly score a journal. Don't be afraid to cover the rest of the journal with a couple of layers of masking tape—just to be safe.

Polish all journals with No. 400 sandpaper, using a shoeshining technique with paper strips.

Somewhere along the line before finally laying that crank to rest in the block, the bearings should be put into place, the crank laid in and the main caps torqued down. Remove them one at a time and set up a dial indicator over each main bearing in turn to determine how much the crank "runs out," if any. The factory allows up to 0.007 inch. A competition-engine crank shouldn't have any more than 0.001-inch run out. Don't let any lop-eared kid tell you that he straightened a large-block Chevy crank in a hydraulic press. This is a surefire method of cracking a crank—although it may be straight after it's cracked. Oddly, cranks are straightened with a hammer by the reputable crankshaft-specialty firms. This is a job for a professional hammer swinger ... not a blacksmith or apprentice carpenter.

According to Bob Joehnck, the No. 2 main journal is the worst offender. If any journal runs out more than 0.001 in., the crank must be straightened. Jere Stahl says that you'll seldom find a Tufftrided crankshaft that is straight as it comes from Chevrolet. They must be straightened almost every time, so don't keep bugging the Chevy dealer to come up with a straight one because he's not likely to be able to.

So, what do you do to the factory high-performance crankshafts to make them right and good for racing use? According the Bob Gillian of Moldex, Chevy has made the process double-simple by providing a super high-quality part to start with. All that you have to do is to make sure that the crankshaft is straight—as we have already explained. If it is not, they can straighten it for you. Magnaflux the crank to check for any hidden flaws—after the straightening routine. Polish the journals to eliminate any rough edges—after the oil holes have been chamfered to eliminate their rough edges. Moldex will check your high-performance crank for straightness, straighten if required, chamfer the holes, polish the journals and Magnaflux it for a relatively low price. See the suppliers list, page 159.

Gillian says that there is no reason whatsoever to shotpeen a Chevy high-performance crank that has been Tufftrided. The shot will merely bounce off of the crank's tough surface and does no good. You might notice a change in the surface color or appearance, but that is all that would happen. If the crank has to be ground undersize, re-Tufftriding will re-establish the rough surface and, more importantly, the surface hardness, at the journals.

Back in the days when guys were finding more horsepower in flathead Fords and six-banger Chevys than the law allowed, they had a lot of problems with the lower ends. In fact, most nearly everything that got the hop-up treatment had lower-end problems of one kind or another. Bearings would score and scuff — cranks would gouge — and even seize. Oil pressure was either of the "tap the gage — I don't think it's working," or "oh my Gawd, I stretched the spring too far" variety. In this era the discovery was made that a groove turned in the main-bearing journal or a 360° groove in the main bearing gave better lubricant distribution to the rods so that they lasted a while longer. Unfortunately, this thinking is still with us and it is one of those holdover pieces of information that is on the ragged edge of being completely useless.

It is true that grooving a non-cross-drilled crankshaft provides a constant source of oil supply to the con-rod bearings, but cross drilling the main journals of the crankshaft eliminates the need for grooves because the connecting rods are then supplied from the oil in the upper bearing half groove, regardless of the position of the crankshaft.

Bottom side of a Wally Booth Pro Stock engine. Stock Chevy four-bolt mains, heavy-duty big-block rods (notice absence of polishing and "romancing"). Mains are not grooved. Radiator petcock allows quick draining of coolant from the block for those fast back-to-back runs.

Cranks, bearings, oil, oiling systems and so on have improved to the point where grooving the crank or both main bearing halves is not only not needed — it is not recommended nor used by Chevrolet! Any engine builder who insists on grooving a crankshaft these days is reducing the life of the engines which he builds. GM's best engineering efforts are incorporated in the stock parts, so put those old unsound ideas out of your head and take your gal out to dinner with the money you save by not grooving the crankshaft and not buying "special" bearing sets with 360° grooves.

A couple of points — or even three — need to be made. And, these are hardly ever thought of when this subject comes up in a bench-racing session. When the journal is grooved, you narrow it by the width of the groove — which may not be what you are really after. The groove weakens the crank because it reduces the diameter of all the journals which are thus machined. Third, and most important, even a very narrow groove in either the journal or the lower main-bearing insert reduces the bearing's load-carrying capabilities by at least 60%. That wasn't what you wanted, *was it?*

Fluid-dynamics sections of fluid-engineering handbooks tell us that the strength of an oil film across a bearing depends greatly on the width of the bearing. The pistons are constantly trying to push the crank out of the bottom of the block and the bearings and the oil film must also support the static weight of the crank and everything which is attached to it. Why would you want to reduce the strength of the all-important oil film with a grooved bearing or crank?

You really only need enough brainpower to heat your hair oil to see that such extra effort and expense to reduce the reliability of your engine borders on stupidity! If you forget these three reasons, fine. Remember this one. There are great bunches of guys running large-block Chevys — from Can Am on down— with non-grooved and non-extra-counterweighted Chevrolet cranks *without failures*. And a lot of them are using stock oil pumps, too.

If you are experiencing chronic crank or bearing failures, there are a number of areas in which to search before turning to crank grooving, which will not ever solve the most common problems which could occur. For instance — is the crank bent? Is the block straight? Were the rod and main clearances correct — or did you bother to check? Were the pistons aligned with the rods so that the bearings would not be subjected to abnormal load patterns? Did you blueprint the oil pump, checking that end clearance very carefully? Do you know what the oil pressure was? Did the engine run out of oil pressure consistently in the turns?

Do this. If bearing failure has you in a panic, check for the obvious wrong. If everything seems to be in order—keep looking—but don't even consider grooved lower bearing halves or a grooved crank. Face it. If a Can Am engine can run with a stock Chevy lower end done right, why can't yours?

Bearings used by Chevrolet are of the correct type, size and have the oil holes and grooves located and sized to lubricate the load-bearing surfaces and to distribute oil to the critical areas. Federal-Mogul's excellent booklet, "Bearing Down," points out that you should install bearings "as is" without enlarging or adding oil grooves which can disrupt the necessary oil film, cause a loss of oil pressure and reduce the fatigue strength of the bearing because of the reduction in total square inches of bearing area. And, incidentally, you'll notice in the heavy-duty parts list that the bearings Chevy supplies for the exotic alloy-block ZL-1 and Can Am engines are the exact same parts that you can buy at your dealer's for your cast-iron engine.

Now that you are convinced, don't become unconvinced by the parts that some auto parts counterman tries to sell you. Open the main-bearing set right there and examine it. If the lower main-bearing halves are grooved, give the set back. If the store does not have a correct set, go back to the Chevy dealer and have him order the right ones if he does not already have them in stock. Don't be confused. Several of the major bearing manufacturers offer 360° grooved main-bearing sets for the big block Chevy. Nearly all of these manufacturers, although they knew better, finally "gave in" and produced these sets to meet the demands of the buying mechanics who insisted that they had to have full-circle-grooved bearings. Some of these makers have tried to justify their position by saying that the bearings thus run cooler — and then they have the gall to advise you to run heavier oil, such as SAE 40 — and a heavy-duty high-volume pump. The pump costs you more money to buy and then steals HP to operate it — HP that you could use at the rear wheels.

So, just because a large percentage of the buying public insists on buying the wrong parts because of a "fix" that worked 20 years ago does not give you any excuse to do the same thing. Now you know why Chevy does not sell grooved cranks or main-bearing sets with grooves in the lower halves, so don't be misled.

Rather than joining the ranks of the monkey-see, monkey-do mechanics, pay attention to what has been proved to work in the big-block Chevy. When you see a whole pile of expensive drag-race engines stacked up at Edelbrock Equipment Co., or Bob Joehnck's, or Bill King's Engine Service — check the bottom ends when you see them looking over an engine. Whether you are peering at engines of drag greats such as Bill Jenkins or Wally Booth — or at one of Denny Hulme's McLaren Can Am engines — you'll see cranks without extra counterweights. Note that the main journals of the cranks are stock — with no grooves added. The main caps will hold stock non-grooved main-bearing inserts, too. When you see that some of the engines, especially those of the drag racers, are built with non-polished rods with only pin-oil holes added—you get the idea that their builders once travelled "the road to romance," but gave up on the trick stuff to settle down with stock-Chevrolet lower ends.

High-performance cranks are cross-drilled by the factory so that one side of the crank-journal drilling — or the other — is always exposed to the oil pressure in the top bearing shell groove. The rods are thus assured a continuous oil supply. Look at the main journals of your crank. If it has holes that go straight through each of the

Welding rods shoved through the main journal and into a rod journal on this hi-perf crank show the cross-drilled design which ensures that con-rod bearings are always oiled — regardless of crankshaft position.

Junior Johnson's big-block-powered Monte Carlo humbled NASCAR racers throughout the 1972 season.

Arrows indicate the grooves which you DO NOT WANT under any circumstances! When any crankshaft outfit or engine builder tries to tell you that grooves are the hot setup and only way to go for the big-block—and then presses you to buy the extra counterweights as well—cough twice and pass. Just buy the stock high-performance Tufftrided crank, check it for straightness, chamfer the holes, polish the journals and get it Magnafluxed. If the crank passes the Magnaflux inspection, you are in business with the finest crank that $$$ can buy.

main journals, you've got the right crank. If not, the holes can be easily added by any machine shop. Just be sure to chamfer the new hole so that it won't plow a groove in the lower bearing half.

Chevrolet has been cross-drilling cranks at the main-bearing journals on their high-performance engines since 1962. With this kind of production experience from GM, it would seem that we might all pay some attention to what's been happening for all these years. They certainly don't do things which aren't necessary — as so many engine builders do — because Chevy engineers are not interested in romance-type modifications. They are in business to build what will run and last at the lowest possible production cost.

Still more about bearings

There's still another old-mechanics fable that persists in spite of all that the bearing companies have done to kill it. This is the one which says that bearings should be "lightly sanded or steel-wooled" before they are installed. This one makes the bearing engineers' hair stand on end. Multiple tens of thousands of dollars are spent in research programs to determine what coating should be put on bearings to resist the abuses of starting up a new engine. Then we "expert" backyard mechanics figure out every possible way to ruin their best efforts. Smart— we're not!

What do you do to install bearings? Clean them in solvent. CLEAN solvent! Dry them carefully with a lint-free rag. Make sure the edges of the rod cap and rod have been chamfered lightly with a fine file. The block and main caps should also be slightly chamfered — right

---

Stroked crankshafts are EXPENSIVE—If the displacement chart on page 100 has you drooling to build a monster motor of say 500 CID or so, consider the cost of the crankshaft. A semi-finished Chevy forging finished out at 4.25-in. stroke will require balancing. A 454 crank which has been internally balanced might turn out to be a far cheaper way to build your engine. Extra inches above 467 CID cost a bundle! Remember that the crank is only the first cost—there are stroker pistons to buy, too—unless you are using a 427 truck block.

at the parting line. Install the DRY bearings in the DRY block, main caps, rod caps and rods. Oil should only be applied to the bearing insert on the load faces, definitely not on the backsides where the inserts seat against the block or rods.

We even found that some mechanics were using Loctite to install rod and main bearings. Bob Joehnck pointed out that Loctite is definitely going to reduce the flow of heat from the bearing into the rod or block because it is a heat barrier which destroys the intimate metal-to-metal contact which helps to allow these parts to live in the engine. Not only that, Loctite is a viscous material which sets up in a fairly thick layer which will reduce the clearance available at your rod or main bearings, Don't use anything on the backside of the bearings — including oil. And, don't chamfer the bearings with a knife, deburring tool or emery cloth. The bearing makers have put the required radius on the insert to match the radius ground into the crank. Save your effort for something that's needed!

Bearing clearances

Although the clearances recommended for these engines may seem to you to be conservative, they must be closely adhered to. Additional clearance might make for a "looser running" engine, but it won't give any more power. The Chevy big-block engines must be set up "tight." They cannot be run loose, as is often done with other high-performance engines, or you will end up with lubrication problems. In checking with dozens of engine builders we could not find any who build engines with larger-than-stock clearances. Increased clearances at the mains and rods automatically increase the oil throw-off and increase the amount of oil which the pump must supply to keep up with what the bearings are throwing off. This can lead to a requirement for a larger pump and larger pumps take more HP to drive.

Additionally, you are creating more problems because the rings were not designed to cope with this additional oil throw-off. And, if you are working with a dry-sumped racing engine, you complicate the problem of getting the extra oil out of the sump and back into the supply tank — a problem which is bad enough without making it worse.

Extra side clearance at the rods is not wanted because this increases the hammering effect of the pins on the pin retainers, thereby adding to the possibility of these items failing.

Although we've warned you about this in other places in the book, we'll say it again. *Don't try any trick clearances in these engines.* Build yours with the correct lifters, rocker arms, cam bearings and clearances to require a MINIMUM oil flow. This greatly aids the functioning of the rings and the oiling system. Ron Hutter of Trick Stuff Engineering comments on the lower end as follows, "The bottom end of the big-block Chevrolet is *very durable* and the stock oiling system is completely adequate if the L-88 pump and deep-sump pan are used. the only critical items are the rod and main-bearing clearances and the chamfer on the oil holes. The single most important thing to me seems to be the willingness of the individual to use good practice and common sense and be willing to take the time to pay attention to detail and cleanliness."

In the event you have a crank with high-limit journal diameters, Chevy offers a line of heavy duty bearings with 0.001-inch larger ID to INCREASE clearance to the recommended specs without having to undergrind a new crank. Get the part numbers from the heavy-duty parts list on page 150.

Pen tip indicates radius which should be present on all crank journals. Always check bearing inserts to make sure that they do not touch this radius. Note the chamfered and polished oil-hole edges.

Stock bearings are the hot tip for any rat motor. Grooved halves are uppers which fit in the block, solid halves are lowers which go in the caps. These bearings, in combination with the cross-drilled hi-perf crank, eliminate any need for trick rework of the lower end. Don't buy main sets with fully grooved bearings. Don't ever groove a rat crank. Low-performance cranks can be cross-drilled at the main journals.

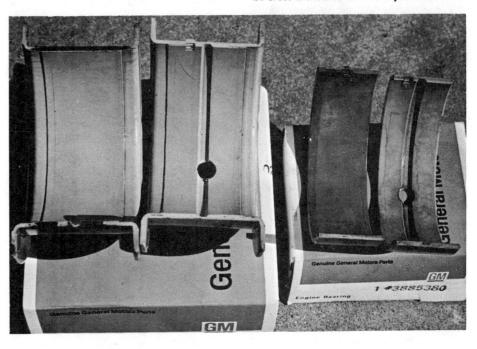

Connecting Rods

If there has ever been a chronic weak point in the big-block Chevrolet engine, it is the connecting rod. Therefore, special attention should be given to the rods for any of the engines which will see anything more demanding than normal street driving. Because Chevrolet is well aware of the difficulties encountered in the past with 3/8-inch-bolt rods, they've just about solved all of the problems with rod 3969804 (was 3959187). It has ground split surfaces, is selected for optimum hardness, run through a number of inspection procedures, Magnafluxed, and equipped with 7/16-inch-diameter super-duty boron-steel bolts with ground shanks. The floating-pin rod is highly recommended as the "hot tip" for building a high-quality heavy-duty engine from parts which are straight out of the Chevrolet boxes. These rods have been used since mid-68 in all L-88s and ZL-1 engines.

The same forging, made to less-exacting specifications is available in a pressed-pin rod for the LS-6 425/450/460 HP. These rods, P/N 3963552, have 7/16-inch knurled-shank bolts. With a lot of hand work, this part can be modified into an excellent racing rod as described in the following paragraphs, but make certain that you add the 3969864 boron-steel bolts. Knurled-shank bolts are definitely *not* recommended for heavy-duty use. Nuts, 3942410, are the same for rod 3969804 with ground-shank bolts and rod 3963552 with knurled-shank bolts.

For an engine destined for heavy-duty action, connecting rods should be reworked to improve pin oiling by drilling a *single* 1/8- or 5/32-in. hole through the *top* of the rod boss. Chamfer the hole with a 1/4-in. drill and deburr the pin bore before honing the boss to provide 0.0005 to 0.0007-inch pin clearance. Piston pins should be individually fit into the rods. Non-floating connecting rods can be reworked in this way to allow their use with full-floating pins. Many builders add this very important oiling hole to the 3969804 rods and then follow through with all of the connecting-rod improvements described in the following paragraphs and on page 143.

Incidentally, you should not use rods with 3/8-inch bolts unless class rules demand that you do so. Now that the good-guy parts are available with the improved holding power afforded by the 7/16-inch bolts, use them!

Connecting-rod life can be extended by rounding all sharp edges on the I-beam section of the rod and removing all excess flash at the forging parting line. You don't need to polish the rod, but getting the flashing off opens up any hiding places for cracks so that the Magnaflux will allow you to find them. All grinding must run lengthwise. Round all of the sharp edges around the rod-bolt head and nut seats and put that hand grinder to work smoothing out any nicks in the radius of the bolt and nut seats. Magnaflux the rods and caps and bolts. If you have access to a hardness tester, check the rod bolts and nuts. Boron-steel bolts should measure 40–48 Rockwell "C," nuts should come up to 32–38 "C." If not, pitch them in the trash.

Checking hardness on a rod bolt or nut is not the easiest thing in the world — but try this procedure. Lightly grind the Lubrite coating off both ends of the bolts and check for hardness on either end. Nuts are even more difficult to test for hardness. Lightly grind both the top and bottom surfaces and CAREFULLY center the point of the hardness tester on the ground surface. If the impression is slightly off center, the nut will indicate softer than it actually is. After you've gone through this with all eight rods, you'll probably wish you'd never seen a connecting rod, so now's a good time to send them out to be shotpeened.

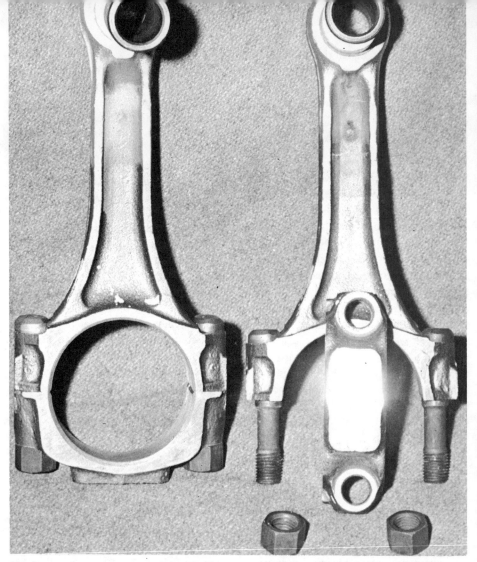

The factory's good-guy rods, P/N 3969804. Balancing lugs are provided at each end of the rod. Cap is secured with 7/16-inch boron-steel bolts with ground shanks. Dull finish at the small end of the rods indicates that rod "eyes" have been coplated with a tin/lead coating. This provides a non-galling bearing surface which permits running the steel piston pin on the steel surface of the rod eye. These rods can be further modified as described in the text. A similar rod for pressed-in pins, 3963552, uses the same forging.

Take the rods apart first. Specify that the pin openings be masked off before peening. Tell the man that you want the rods and caps peened to 0.012 to 0.015-inch Allmen "A" arc height using No. 230 steel shot. *Don't shotpeen the rod bolts or nuts.* Have the rods resized after peening. Incidentally, it's low-cost insurance to use new, Magnafluxed rod bolts and nuts at each teardown.

If all of this care and treatment of connecting rods seems a waste of time and money, keep in mind the fact that rods usually come apart when the engine is winding near its peak. When this occurs, the rod always wipes out other parts on its way out through the side of the block. The intake manifold and water pump are often the only salvageable parts from a big-block which loses a rod.

You should be aware that the crankshafts are factory balanced for use with a particular rod type. That is, engines supplied with 3/8-inch-bolt rods have cranks balanced for use with those rods. If you switch rods, the crank must be balanced.

Here's how Bill King makes his "Superized" rods which have been used in many Can Am and pro-stock engines:

1. Sides of rods are ground for clearance so that sides are at 90° to the bore and parallel with each other.
2. Rod is taken apart for polishing and for drilling the two holes required to lubricate the pin. Nut and bolt seats are radius-ground around the edges so that there are no sharp corners from which a crack could start.
3. Rods are Magnafluxed. Any that don't pass are scrapped. Rods are then shot-peened.
4. Parting face of rod and cap is accurately ground.
5. New boron-steel bolts are center-drilled at each end to provide for location of the stretch-measuring tool.
6. Bolts are installed and tightened to provide correct stretch. This often requires more than specification torque.
7. Rod is sized at the big end and at the small end. All rods are identically sized for length.

King's rod-stretch measurer operates with center indents in bolts to indicate stretch. 1:1 adapter connects to torque wrench. Tool costs racers $84 with indicator and a bridge to use the indicator for deck-height measurements.

Aluminum-tubing "chopstick" is slipped over upper rod bolt when installing the rod and piston assembly. Rod allows guiding the rod over the crank journal so cylinder wall and journal are not scratched. Put one of these in your tool box!

THIS IS THE RIGHT WAY TO OIL THE WRIST PIN. It was once thought to be the right way, then Chevy changed their minds—and now it is the RIGHT WAY—again!

NOTE: — Lengthwise cracks in the rod surface may become apparent after grinding. These forging laps or surface cracks occur when forging flash is sheared off. Engineers we talked to at Chevrolet say they've NEVER seen a con-rod failure due to these cracks. Despite that, many reputable engine builders reject rods with these small lengthwise cracks.

Changing rod bolts

Changing rod bolts is not just a simple matter of hammering out the old and banging in the new ones. If you follow this course of action, you'll probably distort the rod so that the bearing wipes unevenly—and wipes itself out on the rod journal. The rods should *always* be reconditioned any time new bolts are installed. There are no exceptions.

After reconditioning, save the full-torque installation until you are putting the engine together for the last time before running it. That way you can check the bolt for stretch with a micrometer prior to torquing it. Make sure that the rod does not stretch more than 0.005 inch beyond the recommended maximum. A "stretchy" bolt is not wanted in any engine because one of those can stretch in use and cause a catastrophic failure which will demolish the engine and your bank account in one fell swoop. At the least, a stretched bolt could allow the rod bearing to spin, smoking the crank in the process. If you encounter a "stretchy" bolt, take it out immediately and destroy the threads with a hammer blow before pitching it in the trash. That way no one will "save it" for you.

Bill King sells a dial-gage tool which allows you to measure the rod-bolt stretch as you torque the nut in place. This makes the connecting-rod installation much easier than trying to tighten each nut a few degrees and then continually rechecking bolt stretch with a micrometer. Remember that stretch is the desired specification, even if you have to apply more than the recommended torque to obtain it.

If you have to clamp a connecting rod in a vise for any reason, be sure to protect the rod from the vise jaws by using pieces of wood or soft metal on each side of the rod. Any vise-jaw marks in the rod or cap forging should be considered stress-raisers which could lead to early rod failure.

Connecting rod side clearances

Do not be tempted to open the con rod side clearances to more than the specified 0.025-inch per pair. That's measured between the two shoved-apart rods. More clearance merely increases the tendency to hammer out the pin retainer by allowing the pin more sidewise movement to act as a battering ram.

If you have to provide increased clearance, the usual way that this is done is to make a mandrel which is about 0.002-inch larger than the rod bearing ID. This mandrel is chucked in a lathe and the rod is clamped over it. The side of the rod which will be in the center of the rod pair should be turned off, taking care that you do not increase the clearance beyond the recommended maximum of 0.025 inch. By taking the material from the surface which will rub against the other rod, you leave the clearance for the crank radius untouched. That way there is little chance that the rod chamfer at the cheek edge will ever touch the radius between the cheek and the journal. If the clearance has to be increased, take one half of the requirement off of each rod. Example: A rod pair with 0.010-inch side clearance should be increased to the 0.015-inch specification by taking 0.0025 inch off of the center face of each rod.

When you have installed the inserts and bolted the rod cap onto the rod and torqued the bolts, you can check the rod insert inside diameter. Then you can compare this against the diameter of the rod journal to see how much clearance that you have. Measurements of the rod insert ID can be made with a snap gage which is subsequently checked with an outside micrometer. Or, an inside micrometer can be used if it, too, is subsequently measured with an outside micrometer.

If you are going to build engines professionally, then plan to invest in the bore gages and ring standards that are required for the types of engines that you will be building. Although these tools are arm-and-a-leg expensive, they are definitely worth the extra cost and peace of mind that they'll give you in providing measurements that are undeniably accurate and much faster than any other type of device.

Don't rely on Plasti-Gage measuring strips to tell you what kind of clearances that are in the engines that you build. This material does not always provide an accurate indication of the actual clearance. The only way to be sure is to measure ID and compare it against journal diameter. This is true for both rod and main bearings, of course.

When checking rod side clearance, make sure the rod will contact the side of the crank journal and not hang up on the journal fillet. If the rod contacts the radius, it will cause immediate failure of the bearing or at best severe galling of the rod bore and journal fillet. Also check to see that the pin end of the rod does not touch the wrist pin boss on either side of the piston.

Check side clearance between each set of rods as you make a trial assembly. Steel rods should not exceed 0.025 inch clearance, aluminum rods not over 0.030 inch.

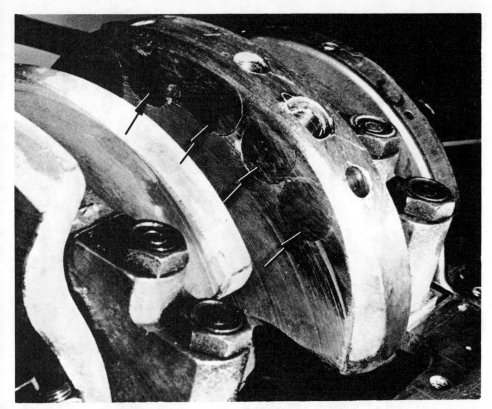

Arrows indicate additional weight which has been added to 454 crank to provide added internal balancing so that non-counterbalanced flywheel and crankshaft harmonic balancer can be used. This is an expensive racing-only modification which should not be considered for a street machine.

The wrong way to do it! Any clamping of a connecting rod in a vise should be done between two boards—one on either side of the rod. The clamp should be made around the large end of the rod, including as much of the cap as possible.

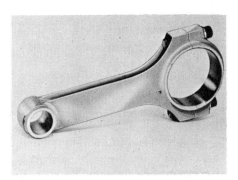

Carrillo rods are acknowledged to be the ultimate for racing engines. Eight will eat a $1000 bill.

Aluminum rods

Aluminum rods? Aluminum rods?? Aluminum rods??? Pro-stockers that you've admired use the stock GM forged-steel rods. Can Am racing-engine builders laugh when anyone mentions any rods other than GM's finest or Carillo's custom-made forged-steel rods.

Here we have another of those "romance items" that are ever present in the speed-equipment business. It sounds so neat to say that an engine has aluminum rods that the buyers of these rods have usually overlooked the only reason for the existence of these parts. They are "shock absorbers" which help to soak up the hammer-like blows of occasional detonation which inevitably occurs in a nitromethane/methanol-burning drag engine. There is absolutely no reason to include aluminum rods in a gasoline-burning engine—and a lot of good reasons not to run them. They have a short life because aluminum work-hardens with use. The block has to be notched to provide clearance. And the super-large cross sections required to provide load-bearing strength equivalent to forged steel makes the aluminum rods weigh darned near as much as steel rods. This is another area where the money that you save by not buying the wrong parts can be used elsewhere. Maybe that money you were planning to spend on aluminum rods could be used to buy another set of rear end gears or a different set of wheels or slicks to get better 1/4-mile times.

Perhaps the explanation of why aluminum rods are so widely used is that they are a part of racing "romance." And, *they are cheap* compared to the price for a set of fully prepared Chevy rods or $1000 for the ultimate H-Beam Carrillo rods. If the aluminum rods are run for a limited time and then replaced, they could be the correct choice for some engines. Living with the wonder of when one will quit would cause us to recommend using the LS-6 rods modified for full-floating pins and the LS-7 bolts drop right in. The price and quality are hard to beat and you know that they are plenty strong, even if you skip the clean-up and shotpeen deal.

Kent-Moore tool for installing harmonic damper. Photo from Chevrolet Overhaul Manual shows how it is used.

Harmonic Balancers

After seeing a harmonic balancer ring either explode or walk off the rubber mounting to its hub, we wondered whether there was any reason to become concerned about these items as a device requiring protection such as we give the flywheel with a scatter shield. According to Chevy's Bill Howell, who has run a flock of endurance tests on the engines, such behavior is not common. It may be your warning that the crankshaft is broken because the balancer soaks up torsional vibrations—which increase when a crank is cracked or badly out of balance, as in the case of a poorly done welded stroker. Such engines sometimes try to tell you about their problems ahead of time by shedding belts and/or pulleys and the harmonic-balancer damper ring. So, be sure to put that balancer in the box of things you were taking to be Magnafluxed.

Assembling any pre-69 big block requires a trick to put on the harmonic balancer and timing cover. These parts are not dowelled to cause automatic alignment in the earlier engines, so careful assembly is essential. If you insist on bolting the cover to the block and then install the balancer, chances are 1,000 to 1 in favor of your ending up with an oil leak caused by seal damage. What to do? Put the cover on loosely with its gasket. Grease the seal lip with high-temp Lubriplate or other suitable grease and install the balancer on the crank snout with the correct tool. The

Next time put the balancer on with the correct installation tool instead of driving it on with a hammer. Hammering the balancer onto the crank is a sure way to destroy the rubber bonding between the balancer center and rim. Note that braided-steel lines will not withstand this kind of shrapnel—nor did the fiber-glass bottom of the boat. Fortunately, the severed line was carrying water instead of fuel!

balancer hub centers the seal so you can seat the cover onto the block by tightening the bolts. We are not saying that this does not require care, it's just that the method usually avoids seal damage. There is a tool for this, but there's no need to buy any more special tools than we have to.

Installation of the timing cover is simplified on 69 and later engines with dowels in the block and matching holes in the cover. You'd be well advised to install similar dowels in your pre-69 block with matching holes in the cover. If the cover is correctly installed, drill through the cover and into the block, then Loctite the dowels in the block after removing the cover.

Use the correct installation tool – Kent-Moore J-21058 – to install the balancer. This tool screws into the crank snout and the balancer is drawn into place with a nut which works against a thrust bearing. Do not, regardless of your hurry, ever try to install the balancer by driving it on with a hammer.

The Shop Manual says, "CAUTION: The inertia-weight section of the torsional damper is assembled to the hub with a rubber-type material. The installation procedure (with correct tool) must be followed or movement of the inertia-weight section on the hub will destroy the tuning of the torsional damper."

And, don't follow the time-honored racers' scheme of heating the hub before sliding the balancer on the oiled crank. This leads to instant seal failure if the hub is hot enough. The installation of the balancer is another of the instances where the right way and the easy way turn out to be the same ones.

A few words about head gaskets — are bigger bolts really necessary?

Some builders insist that larger head bolts are needed for the big block. So, we made photos and detailed instructions for installing them — and we were ready to tell you how to do this in the book. But, in asking further, we found that the use of this expensive idea is pretty well limited to a few shops. And, we couldn't document any real need for the operation.

Bob Joehnck told us, "If you use bigger bolts you have a problem because then the gasket has to be modified and the head holes have to be opened up, too." Joehnck uses chrome-moly bolts and makes double-sure that the lower row of holes on each side of the block are tapped all the way into the water jacket. No extra bosses (ZL-1 style) are added to cast-iron blocks. Bolt threads are coated with Perma-Tex Hytack, which is also applied in a liberal "freckle coat" to an 8180 PT Fel-Pro Perma-Torque gasket. Even though the sandwich-style gasket is called Perma-Torque, Joehnck insists that the heads must be retorqued after the engine has been leaned on hard enough to warm up thoroughly and produce good power. This is obviously a fine line, so the best thing to do is to make a couple of half-hearted passes with the car before shutting it off and letting it cool prior to retorquing the bolts or nuts. If he uses studs, these are installed in the block with mild Loctite. Donovan Engineering adds grooves for copper armature wire. 0.030 to 0.040-inch diameter wire extends 0.007 to 0.010-inch out of the head. Use of the grooves and the armature wire eliminates any need for extra capscrews or studs, according to Joehnck.

We'd suggest that you put your engine together without any expensive grooves and only add these if you are unable to hold the head gaskets with the stock arrangement.

Bill King, on the other hand, prefers Victorcore gaskets supplied by Chevy as 3969865 (listed in HD Parts List). This red composition gasket compresses to 0.040-inch thick when installed. Stainless steel beads around each cylinder retain the pressure. King claims that these gaskets are more reliable than the Victor gaskets sold in a parts store (0.032-inch thick). Although the factory manuals state that the gaskets should be installed dry, we could not find any high-performance people who were willing to do it. King suggested that you use Permatex No. 2 — that dark-brown sticky, gooey junk — at the top of the gasket on both sides so that there will be a little extra holding power in this area. If this is done, you should not have any problem with blown gaskets, even though no extra studs are installed as on a ZL-1. If an engine is to be supercharged or run on nitromethane fuels, he feels that a pair of extra bolts on each side should be added to the cast-iron blocks.

If the aluminum heads are used, hardened steel washers (see parts list) must be used under the capscrews. Make certain that all of the threads on all of the capscrews are clean. Use a non-hardening sealer on the threaded portion of these because they live in water. Permatex No. 2 works fine, or get a can of GM's sealer 1050805 in a spray can — or 1050026 with a brush. Use alcohol to clean up afterwards. In torquing down a head, follow the chart and take plenty of time to do the job. Go around each head 10 or 12 times to build up to the final reading in the specs.

Dino Fry says that a lot of builders create unneeded work for themselves because they are unaware that a valve-grind gasket set exists with the good Victorcore gaskets and with intake gaskets having the exhaust heat riser openings blocked. This is 3974218.

Something is missing —

A number of specifics have been left out of this blueprinting chapter because some portions of the engine are so important that they require detailed coverage in separate chapters. These include pistons, cylinder heads, camshaft and valve-train components, oiling system, ignition, exhaust, flywheel and clutch. Much of the special detail for assembling the big-block engine in high-performance form has been carefully detailed in these special sections of the book. When these recommendations are combined with an intelligent application of the information in this section and in the Service and Overhaul Manuals, you can expect to build a winning combination. Bolt torque and clearance specifications have also been included in this book for your guidance.

A big screwdriver or a bar can be used to wedge the crankshaft fore-and-aft in the block to measure end clearance with a dial indicator. Factory technicians use a big lead mallet to move the crank back and forth against the thrust surfaces when checking end clearances. The same tool can be used to line up the rear main cap and bearing. Whacko!

The necessary tools for installing pistons correctly. You'll also need a ring installer. Oil piston skirt and rings lightly and line up the ring gaps according to the page 108 diagram. Any oil smeared on the piston skirts or in the cylinder bores must be spread with your hand. Avoid using shop rags around the engine during assembly because they leave unwanted lint which can cause trouble.

Variations on a powerful theme
the big-block engine family

KNOW WHAT YOU HAVE!

In assembling this chart we found owners with models having equipment different from that which is described. Further checking revealed that a dealer has the option of using whatever parts he has on hand to make repairs, especially in a warranty or pre-delivery situation. Thus, it's possible for a two-bolt engine to be replaced with a four-bolter and vice-versa if there is a parts shortage, strike or whatever. However, if you are racing in a class where rules are strict, it is up to you to make sure that you are racing what you think you have. Tech inspectors have heard all kinds of funny and phony stories about how an engine was built in a certain manner. You have a poor chance of talking one of them into accepting a weird combination.

Buy any size engine you want for a few cents at your Chevy parts counter. This should warn you that you'd better measure the bore and stroke when you are buying an engine. It could have been decal-ized instead of equipped with the displacement that you think it has.

BIG-BLOCK CHEVROLET VERSIONS
Grouped by displacement and Regular Production Option (RPO) Number, and in order of increasing performance.

396 CU. IN. DISPLACEMENT ENGINES (4.094 x 3.76 Bore & Stroke)

L-66 1969 only — 265 BHP 2-Bbl carburetor, low-lift hydraulic camshaft, small-port cast-iron heads, 9:1 CR, 2-bolt mains, nodular-iron crankshaft.

L-35 1965-69 — 325 BHP Holley or Q-Jet 4-Bbl carburetor (Q-Jet only after 1967), low-lift hydraulic cam, small-port cast-iron heads, 10.25:1 CR, 2-bolt mains, crankshafts: forged 1965-67; nodular iron 1968-69.

L-34 1966-69 — 360/350 BHP Holley 4-Bbl carburetor until 1968, then Q-Jet, 1968 and 69, high-lift hydraulic cam, small-port cast-iron heads, 10.25:1 CR, 2-bolt mains (some 1966 Chevelles had 4-bolt mains), forged crankshaft.

L-78 1965-69 — 375 BHP (rated 425 BHP in 1965 Corvette only) Holley 800 CFM 4-Bbl carburetor, mechanical-lifter cam, large-port cast-iron heads, 11:1 CR, 4-bolt mains, forged Tufftrided crankshaft, forged pistons. Sold in Chevelle, Chevy II and Camaro.

L-89 1968-69 — 375 BHP Limited edition of L-78 engine produced with non-open-chambered aluminum cylinder heads in Chevy II and Camaro.

402 CU. IN. DISPLACEMENT ENGINES (4.126 x 3.76 Bore & Stroke)

LS-3 1970-71 — 330 BHP (300 BHP 1971) Q-Jet 4-Bbl carburetor, low-lift hydraulic cam, small-port cast-iron heads, 10.25:1 CR (1970) and 8.5:1 CR (1971), 2-bolt mains, nodular-iron crankshaft.

LS-3 1972 — 240 HP w/dual exhaust, 210 HP with single exhaust, other specs same as 1970-71 LS-3.

L-34 — 350 BHP Q-Jet 4-Bbl carburetor, high-lift hydraulic cam, small-port cast-iron heads, 10.25:1 CR, 2-bolt mains, forged crank.

L-78 — 375 BHP Identical with L-78 396 except for bore size.

427 CU. IN. DISPLACEMENT ENGINES (4.251 x 3.76 Bore & Stroke)

LS-1 1969 only — 335 BHP Q-Jet 4-Bbl carburetor, hydraulic camshaft, small-port cast-iron heads, 10.25:1 CR, 2-bolt mains, nodular-iron crank.

L-36 1966-69 — 390/385 BHP Holley or Q-Jet 4-Bbl carburetor, high-lift hydraulic camshaft, small-port cast-iron heads, 10.25:1 CR, 2-bolt mains, forged crank.

L-68 1967-69 Corvette — 400 BHP Identical with L-36 except for 3 Holley 2-Bbl carburetors on small-port aluminum intake manifold.

L-72 1966-68 — 425 BHP Holley 800 CFM 4-Bbl carburetor, mechanical camshaft, big-port cast-iron heads, 11:1 CR, 4-bolt mains, forged Tufftrided crank, forged pistons, aluminum high-rise intake manifold.

L-71 1967-69 Corvette — 435/425 BHP Identical with L-72 except for 3 Holley 2-Bbl carburetors on aluminum manifold.

L-89 1968 Corvette — Similar to Corvette L-71 except for aluminum cylinder heads (non-open-chamber-type).

L-88 1967-69 Corvette	430 BHP	Holley 850 CFM 4-Bbl carburetor, high-lift mechanical cam, aluminum cylinder heads (open chamber in 1969), 12.5:1 CR, 4-bolt mains, forged Tufftrided crank, forged pistons, floating wrist pins, 1/16" compression rings, 7/16" push rods, open-plenum intake manifold.
ZL-1 1969-70 Corvette	430 BHP	Identical features to the L-88 engine except all-aluminum cylinder block.

454 CU. IN. DISPLACEMENT (4.251 x 4.00 Bore & Stroke)

LS-4 1970 only	345 BHP	Q-Jet 4-Bbl carburetor, low-lift hydraulic camshaft, cast-iron small-port cylinder heads, 10.25:1 CR, 4-bolt mains, forged crankshaft.
LS-5 1970	360/390 BHP	Q-Jet 4-Bbl carburetor, high-lift hydraulic camshaft, small-port cast-iron heads, 10.25:1 CR, 4-bolt mains, forged crankshaft.
LS-5 1971	365 BHP	Specs as per 1970 LS-5, except for small-port open-chamber cast-iron heads with 8.5:1 CR.
LS-5 1972	270 HP with dual exhaust, 230 HP with single exhaust, other specs same as 1971 LS-5.	
LS-6 1970	460/450 BHP	Holley 800 CFM 4-Bbl carburetor, mechanical cam, cast-iron large-port cylinder heads, 11:1 CR, 4-bolt mains, forged Tufftrided crankshaft, forged pistons with pressed-in pins.
LS-6 1971 Corvette	425 BHP	Holley 800 CFM 4-Bbl carburetor, mechanical cam, aluminum large-port open-chamber cylinder heads, 9.0:1 CR, all other specs as 1970 LS-6. Cast-iron large-port open-chamber heads available as service parts.
LS-7 1970 Corvette	465 BHP	Never released for production by Chevrolet. Consisted of Holley 850 CFM 4-Bbl carburetor, mechanical cam, large-port aluminum heads (open-chamber), 12.5:1 CR, 4-bolt mains, forged Tufftrided crankshaft, forged pistons, full-floating wrist pins, 1/16" compression rings, 7/16" push rods, open-plenum intake manifold. May be assembled from available Chevy service parts.

GENERAL COMMENTS

Power Ratings

1972 power ratings were reduced because GM changed to an SAE 85°F correction factor rather than the 60°F factor previously used for the gross power rating method.

Valve Sizes

All small-port cast-iron heads use 2.07 dia. intake and 1.725 dia. exhaust valves.

All large-port cast-iron heads 1965-1970 use 2.195 dia. intake and 1.725 exhaust valves.

1971 large-port open-chamber cast-iron heads use 2.195 dia. intake and 1.885 dia. exhaust valves.

All non-open-chamber aluminum cylinder heads 1967-1970 use 2.195 dia. intake and 1.84 dia. exhaust valves.

1969-1971 open-chamber aluminum heads use 2.195 dia. intake and 1.88 dia. exhaust valves.

Camshaft Lift and Timing

Low-lift hydraulic camshaft used in all L-35's and L-66 396's, LS-1 and 1970 LS-3: 0.398-inch lift intake and exhaust with 322° duration (including ramps); LS-3 (1971) and LS-4 (1970): 0.398-inch lift intake, 316° duration intake (including ramps) and 0.430-inch lift exhaust, 334° duration exhaust (including ramps).

High-lift hydraulic camshafts used in all L-34, L-36, L-68, and LS-5: 0.4614-inch lift intake, 350° duration (including ramps) and 0.4800-inch lift exhaust, 352° duration (including ramps).

Mechanical cams used in L-78, L-72, L-71, and LS-6: 0.500-inch lift intake, 316° duration (at lash) and 0.500-inch lift exhaust, 302° duration (at lash). Effective duration is less, see table on page 55.

Engine swappers: look carefully. Corvette left rocker cover with a recess at the rear for power-brake clearance could be really helpful when installing the big-block engine in a tight-fitting chassis. Engine is a 1969 435 HP, 427-incher.

Chaparral-prepared ZL-1 in McLaren Can Am car. Timed fuel injection—usually modified Lucas—is used in most of these racing machines. Sports Car Graphic photo.

Who is "Mark," anyway?

Chevrolet engineers and enthusiasts constantly refer to the big blocks as "Mark" engines. There have actually been several "Marks."

Mark 1 - 348/409 CID series

Mark 2 - 427 CID "Mystery" engine first used at Daytona in 1963

Mark 3 - never released

Mark 4 - 396/402/427/454 — the current Turbo-Jet engines

Heavy-duty parts list

PART NO.	DESCRIPTION	QTY.
	Engine Assembly	
366250	454 CID, RPO LS6. 10.2:1 Compression, 4-bolt-main iron block, 1053 forged-steel crank 3963523, 4340 forged-steel connecting rods w/7/16-in. bolts for pressed pins 3863552, forged pistons, street mechanical camshaft 3863143, open-chamber cast-iron cylinder heads 6260482, aluminum high-rise manifold 3933163, water pump, balancer 3963530, standard flywheel 3993827, (for partial engine use 3981820). (ENGINE SUFFIX XAA)	1
3965774	454 CID, RPO LS7. 12.25:1 Compression, 4-bolt-main iron block, 5140 forged-steel crank w/cross-drilled mains 3963524, 4340 forged-steel connecting rods w/7/16-in. bolts for pressed pins 3863552, forged high-dome pistons, racing mechanical camshaft 3959180, open-chamber cast-iron cylinder heads 6260482, balancer 3963530, standard flywheel 3993827 (does not include water pump or intake manifold). (No partial engine available) (ENGINE SUFFIX XCH)	1
	Gasket Kit, Engine	
	NOTE: Engine gasket kit 476074 together with oil-pan and cylinder-head gasket kits provide all necessary gaskets for complete engine overhaul.	
476074	1965—'86 396—454 CID	1
	Blocks and Partial Engines	
3970699	427-CID Partial Engine. Iron block w/4-bolt mains, 5140 forged-steel crank 3967811, forged-aluminum pistons w/12.5:1 compression, racing mechanical camshaft 3925535, 4340 forged-steel connecting rods w/floating pins and 7/16-in. boron-steel bolts. Requires open-chamber cylinder head (6260482 iron or 14011076 aluminum). From 1968-69 RPO L88 427-CID, 435-HP Corvette.	1
3981820	454-CID Partial Engine. Iron block w/4-bolt mains, 1053 forged-steel crank 3963523, forged-aluminum pistons w/11:1 compression when used w/closed-chamber heads and 10.2:1 w/open-chamber heads, 4340 forged-steel connecting rods with pressed pin and 7/16-in. rod bolts 3963552, street mechanical camshaft 3863143. From 1970 RPO LS6 454-CID, 450-HP Chevelle.	1
3952318	427/454-CID Cast-Aluminum Bare Block, w/4.25-in.-bore iron sleeves, 4-bolt main caps. From 1969 ZL1 Corvette and Camaro.	1
14044809	427/454-CID Cast-Iron Bare Block, w/4.25-in.-bore, 4-bolt main caps. Production Corvette block.	1
14015332	427/454-CID Cast-Iron Bare Block, w/4.25-in. bore, 4-bolt main caps. Production-type 60-series truck block. Note: Deck is 0.400-in. higher than Corvette 427/454 blocks.	1
14044808	427/454/500-CID "Off-Highway" Cast-Iron Bare Block, w/4.25-in. bore, 4-bolt main caps. Siamesed walls can be bored to 4.50 in. Note: Deck is 0.400-in. higher than Corvette 427/454 blocks.	1
	Core Plugs	
10000462	Core Plug, steel, 1-5/8-in. OD. For iron block.	AR
14011003	Core Plug, aluminum, 1-5/8-in. OD. For aluminum block.	AR
3826504	Core Plug, brass, 1-5/8-in. OD.	AR
3743389	Core Plug, steel, 1-3/4-in. OD.	AR
	Main Bearing Caps, Hardware	
14015334	Bearing Cap, crankshaft, 4-bolt production-type. Use on #1, 2, 3, 4 mains. Ends must be machined for proper installation.	AR
14015336	Bearing Cap, crankshaft, 4-bolt production-type. Use on #5 mains. Ends and oil groove must be machined for proper installation.	AR
3859927	Bolt, outer, 4-bolt main, 1/2-13 X 2-39/64-in. long. For all except aluminum block.	10
3909834	Bolt, inner, 1/2-13 X 3.34-in. long. For all except aluminum block.	AR
22510580	Bolt, outer, 1/2-13 X 3.10-in. long. For aluminum-block 427 CID.	AR
3902885	Stud, inner, for oil-baffle tray support, except aluminum block.	4
14011001	Stud, 7/16 X 4.30-in. long.	10
14011002	Stud, 7/16 X 3.00-in. long.	6
3952321	Bolt, inner, 1/2-13 X 4.24-in. long. For aluminum-block 427 CID.	AR
14044866	Nut, 12-point, heavy-duty 4037 steel, 7/16 X 20-in. long.	AR
	Bearing Kits, Main	
347018	Bearing Kit, std., for rear main. Standard high performance and heavy duty.	1
347013	Bearing Kit, std., for front and intermediate mains. Standard high performance and heavy duty.	4
	Engine Cover, Front	
14044815	Front Cover, chrome.	1
	Timing Pointers, Front Cover	
14044821	Pointer, front cover, chrome.	1
	Cylinder Head	
3919839	Cylinder Head, cast-iron. Closed-chamber, 109cc design w/2.19-in. intake and 1.72-in. exhaust. Cannot be used w/open-chamber pistons	2
6260482	Cylinder Head, cast-iron. Open-chamber, 118cc design w/2.19-in. intake and 1.88-in. exhaust.	2
3919838	Cylinder Head, aluminum, 1965—'69 396/427 CID. Closed-chamber, 106.8cc design w/2.19-in. intake and 1.84-in. exhaust. Cannot be used w/open-chamber pistons.	2
14011076	Cylinder Head, aluminum, 1970—'71 427/454 CID. Open-chamber, 118cc design w/2.19-in. intake and 1.88-in. exhaust. C-port type, casting #14011077. T-356 aluminum w/T-6 heat-treat specification.	2
14011004	Cylinder Head, aluminum. Same as 14011076 except solid casting (no water jacket).	2
14044862	Cylinder Head, aluminum, 427/454 CID. Modified, open-chamber, 105cc design w/higher exhaust port, 2.19-in. intake and 1.88-in. exhaust valve, reinforced rocker-arm boss. T-356 aluminum w/T-6 heat-treat specification. Intake flange uses production gaskets.	2
14044861	Cylinder Head, aluminum. Same as 14044861 except no valve seats or guides.	2
3893268	Insert, intake-valve seat, 2.19-in. diameter, for aluminum cylinder head 3919838, 3981070, 14011076, 14044861 or 14044862.	8
3946078	Insert, exhaust-valve seat, 1.84-in. diameter, for aluminum cylinder head 3946072, 3981070, 14011076, 14044861 or 14044862.	8
	Gasket Kit, Cylinder Head	
3969865	Gasket, head. For 427/454 aluminum heads. Composition material w/0.039-in. compressed thickness.	2
3974218	Gasket Kit, head. For 427/454 w/aluminum heads. Includes head gasket 3969865, plus rocker-cover, intake-manifold, water-outlet, distributor, head-pipe, and heat-valve gasket.	1
3976081	Gasket, head. For 427/454 w/4.44-in. bore, composition material w/0.039-in. compressed thickness.	2

PART NO.	DESCRIPTION	QTY.
	Hardware, Cylinder Head	
3877668	Bolt, cylinder head, 7/16-14 X 4.06 in. Cannot be used on aluminum block and aluminum cylinder head 14044861 or 14044862.	24
3877669	Bolt, cylinder head, 7/16-14 X 2.08 in. For cast-iron head.	8
334972	Bolt, cylinder head, 7/16-14 X 5.25 in. For 454-CID aluminum head 14044861 or 14044862.	8
3894327	Dowel, cylinder head, 1/2-in. ID X 5/8-in. OD.	AR
3965763	Stud Kit, cylinder head. Includes 32 studs only. Use w/washer 3899696 and nut 3942410 or 14044866.	AR
14011040	Washer, hardened, 0.45-in. ID X 0.78-in. OD. Use w/stud kit 14014408, main-bearing stud kit 14011039, or aluminum cylinder heads.	AR
3899696	Washer, hardened, 0.45-in. ID X 0.86-in. OD. Use w/stud kit 3965763 and aluminum cylinder heads.	AR
3942410	Nut, cylinder-head, six-point, 7/16-20. 100% magnafluxed steel. Use w/all Chevrolet stud kits.	AR
14044866	Nut, cylinder-head, 7/16-20. Off-highway, 12-point. 100% magnafluxed 4037 steel. Use w/all Chevrolet stud kits.	AR
	Valves, Springs, Shims, Seals, Retainers, Keys	
3969815	Intake Valve, 396/427/454-CID std. stem (2.19-in. head).	8
3879618	Intake Valve, 396/427/454-CID std. stem (2.30-in. head).	8
340286	Intake Valve, 396/427/454-CID std. stem (2.19-in. head). Titanium alloy.	8
3860002	Exhaust Valve, 396/427/454-CID std. stem (1.72-in. head).	8
3879619	Exhaust Valve, 396/427/454-CID std. stem (1.84-in. head).	8
3946077	Exhaust Valve, 396/427/454-CID std. stem (1.88-in. head).	8
340287	Exhaust Valve, 396/427/454-CID std. stem (1.88-in. head). Lightweight, hollow-stem.	8
3893294	Guide, exhaust-valve, for aluminum head.	8
3893295	Guide, intake-valve, for aluminum head.	8
3731058	Shim, (55/64 ID X 1-15/64 OD X 0.030-in. thick).	AR
3875916	Shim, (45/64 ID X 1-31/64 OD X 0.015-in. thick).	AR
3864904	Shim, (45/64 ID X 1-31/64 OD X 0.050-in. thick).	AR
3891521	Shim, (45/64 ID X 1-31/64 OD X 0.065-in. thick).	AR
3970627	Valve Spring, high-performance 396/427/454-CID.	AR
3916164	Valve-Spring Damper, RPO L88, ZL1, LS7 engines. Requires 1.487-in.-OD outer spring.	AR
3989354	Dual Valve Spring and Damper, 396/427/454-CID w/camshaft 3994094. 1.538-in.-OD outer spring.	AR
460527	Seal, valve-stem, oil.	AR
3879613	Cap, valve-spring and seal, steel. Use w/valve spring 3916164.	AR
3989353	Cap, valve-spring, steel. Use w/valve spring 3989354.	AR
3947880	Key, valve-keeper, hardened steel (green).	AR
	Rocker Arm	
6258611	Rocker Arm, steel, ball-and-nut-type.	AR
3959182	Rocker Arm, hardened-steel, w/elongated stud hole for high-lift camshaft. Stamped "H."	AR
	Rocker Cover, Grommet, Retaining Hardware	
14044814	Rocker Cover, chrome, tall design. Will clear rocker-arm-stud girdles. Has Chevrolet bow-tie and word POWER embossed on cover.	2
3894337	Grommet, rubber, 15/16-in. ID X 1-7/32-in. OD. For bow-tie rocker cover 14044814.	AR
14011078	Spring Bar Retainer, narrow-design, for tall or regular rocker covers.	AR
14044820	Spring Bar Retainer, chrome, narrow-design for tall or standard rocker covers.	AR
14082320	Stud, rocker-cover hold-down. Torxhead 1/4-20 X 1-1/4 in.	8
14051876	Nut, rocker-cover hold-down, 1/4-20 in.	8
	Pushrod, Pushrod Guide	
3946067	Pushrod, 8-7/32-in. overall length.	8
3946069	Pushrod, 9-3/16-in. overall length.	8
3942415	Pushrod, intake, 427-CID w/heavy-duty aluminum block. 7/16-in. diameter.	8
3942416	Pushrod, exhaust, 427-CID w/heavy-duty aluminum block. 7/16-in. diameter.	8
3879620	Guide, pushrod. Use with 7/16-in. pushrods 3942415-6	8
	Valve-Adjusting Hardware	
5232762	Ball, rocker-arm, 61/64-in. diameter.	AR
3896648	Nut, rocker-arm.	AR
3921912	Stud, rocker-arm, 7/16-in. diameter.	AR
	Valve Lifter	
5232695	Lifter, mechanical. Piddle-valve design, 2-in. length X 27/32 OD. Use w/heavy-duty valve springs.	16
	Camshaft, Camshaft Bearings	
340282	Camshaft, mechanical. For short-track competition w/single 4-bbl carburetor. 0.560-in. intake lift, 0.560-in. exhaust lift w/1.7:1 rocker ratio. (Ident. 340283). 0.024-in. intake, 0.026-in. exhaust clearance. Use valve spring 3916164.	1
3863143	Camshaft, mechanical. For production 396/427/454. 0.500-in. intake lift, 0.500-in. exhaust lift w/1.7:1 rocker ratio. (Ident. 3863144). 0.028-in. intake, 0.028-in. exhaust clearance. Use valve spring 3970627.	1
3925533	Camshaft, mechanical, gear-driven. For 1967 H27. 0.540-in. intake lift, 0.560-in. exhaust lift w/1.7:1 rocker ratio. (Ident. 3925534). 0.024-in. intake, 0.026-in. exhaust clearance. Use valve spring 3916164.	1
3925535	Camshaft, mechanical. From 1967—'69 427-CID RPO L88. 0.540-in. intake lift, 0.560-in. exhaust lift w/1.7:1 rocker ratio. (Ident. 3925536). 0.024-in. intake, 0.026-in. exhaust clearance. Use valve spring 3916164.	1
3959180	Camshaft, mechanical. From 1969 427-CID RPO ZL1 and 1970 454-CID RPO LS7 competition engines. 0.560-in. intake lift, 0.600-in. exhaust lift w/1.7:1 rocker ratio. (Ident. 3959181). 0.024-in. intake, 0.026-in. exhaust clearance. Use valve spring 3916164.	1
3994094	Camshaft, mechanical. For drag and road race, fuel-injection or large-displacement engines. 0.600-in. intake lift, 0.600-in. exhaust lift w/1.7:1 rocker ratio. (Ident. 3994094). 0.024-in. intake, 0.026-in. exhaust clearance. Use valve spring 3989354.	1
6272989	Camshaft, hydraulic. For marine 454-CID engine, good high performance street cam. 0.500-in. intake lift, 0.500-in. exhaust lift w/1.7:1 rocker ratio. (Ident. 6272988). Use valve spring 3970627.	1
3967808	Bearing, camshaft. (Stamped 1)	1
3967809	Bearing, camshaft. (Stamped 3 and 4)	2
3967810	Bearing, camshaft. (Stamped 2, for position No. 2 and R.R.)	2
3859926	Plug, all except aluminum block. (2-7/32-in. diameter)	1
	Connecting Rod, Bearings, Hardware	
3963552	Rod, 1970—'71 454-CID RPO LS6 and LS7. Pressed-pin design w/7/16-in. bolt 3981092. 6.135-in.-long 4340 steel forging, Magnafluxed and shot-peened (color code white). Torque to 73 ft-lb.	AR
3969804	Rod, 1968—'69 427-CID RPO L88. Floating-pin design w/7/16-in. bolt 3969864 and nut 3942410. 6.135-in.-long 4340 steel forging, Magnafluxed and shot-peened, (color code green). Torque to 67—73 ft-lb or 0.009-in. stretch.	AR
347002	Bearing Kit, connecting-rod, std.	8

PART NO.	DESCRIPTION	QTY.
3981092	Bolt, connecting-rod cap, 1970—'71 454-CID RPO LS6 and LS7. 7/16-20 X 2.28 in. w/knurled shank.	AR
3969864	Bolt, connecting-rod cap, 1968—'69 427-CID RPO L88 and ZL1. 7/16-20 X 2.34-in. boron steel, 100% Magnafluxed and shot-peened.	AR
3942410	Nut, connecting-rod bolt, six-point. 7/16-20 X 7/16-in. thick.	AR
340289	Nut, connecting-rod bolt, heavy-duty 12-point. 7/16-20 in., AMS 6304 steel.	AR
14044866	Nut, connecting-rod bolt, heavy-duty 12-point. 7/16-20 in., 4037 steel.	AR

Piston

PART NO.	DESCRIPTION	QTY.
3959105	Piston Assembly, std., 1968—'69 427-CID RPO L88 and ZL1. Forged-aluminum, open-chamber type w/12.5:1 compression. Floating-pin design. Use pin retainer 3942423. (Ident. 3947886 or 3959108)	8
3959107	Piston Assembly, 0.030-in. OS, 1968-'69 427-CID RPO L88 and ZL1. Forged-aluminum, open-chamber type w/12.5:1 compression. Floating-pin design. Use pin retainer 3942423. (Ident. 3959110)	8
3976013	Piston Assembly, std., 1970 454-CID RPO LS6. Forged-aluminum, closed-chamber type w/11.0:1 compression. Pressed-pin design. (Ident. 3963550 or 3976031). Note: If used w/open-chamber head, compression becomes 10.2:1.	8
3976014	Piston Assembly, std., 1970 454-CID RPO LS7. Forged-aluminum, open-chamber type w/12.25:1 compression. Pressed-pin design. (Ident. 3963551 or 3976032). Use on engine assembly 3965774.	8
3976026	Piston Assembly, 0.030-in. OS, 1970 454-CID RPO LS7. Forged-aluminum, open-chamber type w/12.25:1 compression. Pressed-pin design. (Ident. 3976044). Use on engine assembly 3965774.	8
3981075	Piston Assembly, 0.060-in. OS, 1970 454-CID RPO LS7. Forged-aluminum, open-chamber type w/12.25:1 compression. Pressed-pin design. (Ident. 3981073). Use on engine assembly 3965774	8
6262976	Piston Assembly, std., 1971 454-CID RPO LS6. Forged aluminum w/9.0:1 compression. Pressed-pin design. (Ident. 3994031, 3999295, 6262972 or 6269362-3).	8
6262977	Piston Assembly, std., high-limit, 1971 454-CID RPO LS6. Forged aluminum w/9.0:1 compression. Pressed-pin design. (Ident. 3994032, 6262973 or 6269364).	8
6262979	Piston Assembly, 0.030-in. OS, 1971 454-CID RPO LS6. Forged aluminum w/9.0:1 compression. Pressed-pin design. (Ident. 3994034, 6262975 or 6269366).	8

Retainer, Piston Pin

PART NO.	DESCRIPTION	QTY.
3942423	Retainer, Spirolox design, 1.103-in. OD X 0.050-in. thick. For 1969 427-CID RPO L88.	AR
3964238	Retainer, Spirolox design, 1.103-in. OD X 0.072-in. thick. Use when converting to floating-pin design on 454 CID.	AR

Ring Unit, Piston

PART NO.	DESCRIPTION	QTY.
3993828	Piston-Ring Unit, std., 427/454-CID competition.	8
3993829	Piston-Ring Unit, 0.005-in. OS, 427/454-CID competition.	8
3993980	Piston-Ring Unit, 0.030-in. OS, 427/454-CID competition.	8
3993931	Piston-Ring Unit, 0.060-in. OS, 427/454-CID competition.	8

Crankshaft

PART NO.	DESCRIPTION	QTY.
3963521	Raw Crank Forging, 454-CID, 5140 steel forging. Can be machined to 4-in. stroke. (Ident. 3963521).	1
3967811	Crankshaft, 1968—'69 aluminum-block 427 CID. 3-49/64-in.-stroke, 5140 steel forging w/cross-drilled mains. Nitride finish on journals. 2.1985-in.-rod, 2.748-in.-main journal diameters. (Ident. 7115).	1
3963523	Crankshaft, 1970—'71 454 CID. 4-in. stroke, 1053 steel forging w/cross-drilled mains. Nitride finish on journals. 2.1985-in.-rod, 2.748-in. main-journal diameters. (Ident. 3520 or 7416). Used in engine assembly 366250.	1
3963524	Crankshaft, 1970 454 CID. 4-in. stroke, 5140 steel forging w/cross-drilled mains. Nitride finish on journals. 2.1985-in.-rod, 2.748-in. main-journal diameters. (Ident. 3963521). Used in engine assembly 3965774.	1
3752487	Bearing, pilot, bushing-type. 19/32-in. ID, 1-3/32-in. OD X 3/4 in.	1

Crankshaft Balancer (Damper), Pulley

PART NO.	DESCRIPTION	QTY.
3879623	Balancer, 1967—'69 427 CID, 8-in. OD. Use w/all internal-balance engines.	1
3963530	Balancer, 1970—'71 454 CID, 8-in. OD. Use w/external-balance engines only.	1
3899660	Pulley, crankshaft, cast-iron. Double-deep groove, 6-in. OD.	1
455028	Bolt, crankshaft-balancer, 1/2-20 X 1-1/2 in.	1
3864814	Washer, crankshaft-balancer, 0.52 X 2.06 X 0.28-in. thick.	1

Flywheel, Flexplate

PART NO.	DESCRIPTION	QTY.
361950	Flexplate, auto. trans., 427-CID HD. 14-in. starter ring-gear OD.	1
336717	Flexplate, auto. trans., external-balance 454 CID. 14-in. starter ring gear OD.	1
3991406	Flywheel, man. trans., 427-CID RPO L88, 13 lb nodular iron. Use w/10-1/2-in.-diameter clutch. 12-3/4-in. starter ring-gear OD. (Ident. 3856579).	1
3993827	Flywheel, man. trans., 454-CID RPO LS5 and LS6, external-balance only. Use w/11-in.-diameter clutch. 14-in. starter ring-gear OD. (Ident. 3993457).	1
3963537	Flywheel, man. trans., 454 CID, lightweight. Use w/10-1/2-in.-diameter clutch. 12-3/4-in. starter ring-gear OD. (Ident. 3935411).	1
3701679	Dowel, flywheel, 7/16 X 7/8 in.	1
460583	Ring Gear, 12-3/4-in. flywheel.	1
3991407	Ring Gear, 14-in. flywheel.	1

Timing Chain, Sprockets

PART NO.	DESCRIPTION	QTY.
3860036	Timing Chain, link-type.	1
3860035	Crankshaft Sprocket, steel, 19-tooth link-type. 3-1/32-in. OD.	1
6272965	Crankshaft Gear, for gear-drive cam 3925533.	1
330815	Camshaft Sprocket, nylon, 38-tooth aluminum-link-type.	1
3860086	Camshaft Gear, for gear-drive cam 3925533.	1

Water Pump, Water-Pump Components

PART NO.	DESCRIPTION	QTY.
3770245	Pulley, water-pump, deep two groove. Small hub w/5/8-in. shaft. 7-1/8-in. OD.	1
3720616	Reinforcement, water-pump pulley. For small hub w/5/8-in. shaft.	1
3942992	Pulley, water-pump, deep two groove. Fits large hub, but must be reworked for 3/4-in. shaft.	1
9433722	Drive Belt, fan and water-pump, captured. 34 X 3/8 in.	1
14058915	Water Pump, aluminum, 1966—'70 427/454-CID Corvette.	1

Oil Pan, Baffle

PART NO.	DESCRIPTION	QTY.
3977591	Oil Pan, all 1965—'74 396/427/454-CID Corvette. From RPO L88 and LS7. Used on engine assembly 3965774. Includes baffle 3967854.	1
14081027	Oil Pan, 1970 454-CID RPO LS6. Used on engine assembly 366250.	1
3967854	Baffle Tray, oil-pan. Use w/oil pan 3977591 (requires 4 mounting studs 3902885).	1

Oil Pump, Screen, Regulator Spring

PART NO.	DESCRIPTION	QTY.
3876866	Spring, pressure-regulator, HD. (Light-green stripe).	1
3969870	Pump-and-Screen Assembly, 427/454-CID RPO ZL1 and LS7. 1.3-in. gear. Use w/engine 3965774 and Corvette pan. (Distance from mounting surface to bottom of screen is 4.94 in.)	1

PART NO.	DESCRIPTION	QTY.
475908	Pump-and-Screen Assembly, 396/427/454 CID. Used on RPO LS6. Use w/engine 366250. (Distance from mounting surface to bottom of screen is 4.88 in.)	1
475902	Pump-and-Screen Assembly, 1967—'69 427 CID. (Distance from mounting surface to bottom of screen is 4.94 in.)	1
3955283	Screen Assembly, 396/427/454 CID. (Lowest point of screen assembly on plane parallel to oil-pump attaching surface is 6.05 in.). Note: Weld screen-and-tube assembly to pump for HD usage.	1
3955281	Screen Assembly, 396/427/454 CID. (Lowest point of screen assembly on plane parallel to oil-pump attaching surface is 4.88 in.). Note: Weld screen-and-tube assembly to pump for HD usage.	1
6269895	Screen Assembly, 396/427/454 CID. (Lowest point of screen assembly on plane parallel to oil-pump attaching surface is 4.94 in. Note: Weld screen-and-tube assembly to pump for HD usage.	1

Breather, Oil Filler

PART NO.	DESCRIPTION	QTY.
6421868	Oil Cap and Breather, push-in type for HD chrome rocker covers.	1

Adapter, Oil Coolers

PART NO.	DESCRIPTION	QTY.
340258	Adapter, oil-cooler and filter. Allows installation of oil-cooler lines w/side outlets.	1
5575416	Valve Assembly, oil-cooler bypass. Used on 427/454 CID w/4-bolt-main block.	1
340258	Adapter, oil-cooler and filter. Allows installation of oil-cooler lines w/forward outlets.	1
3951644	Bolt, adapter, 5/16-18 X 1-1/8 in.	2
326100	Gasket, oil-cooler adapter and filter.	1
14015353	Seal, oil-cooler adapter and filter.	1
3853870	Connector, oil filter for screw-on filter (required with oil-filter adapter and cooler).	1

Starter Motor, Housing, Hardware

PART NO.	DESCRIPTION	QTY.
1108789	Starter Motor, HD for 12-3/4-in. flywheel. Use w/flywheel 3963537 or 3991406.	1
1108400	Starter Motor, for 14-in. flywheel. Use w/flywheel 361950, 336717 or 3993827.	1
14057099	Bolt, starter-mounting, long w/special shoulder, 3/8 X 4-21/32 in.	1
14057098	Bolt, starter-mounting, short w/special shoulder, 3/8 X 1-27/32 in.	1
3733289	Bolt, starter-mounting, 1966—'68 427 CID, 3/8-16 X 3-5/8 in.	1
354353	Brace, starter-motor, front support.	1
1968122	Starter Housing. Use on starter motor 1108381 or 1108789 for 12-3/4-in. flywheel ring.	1
1984098	Starter Housing. Use on starter motor 1118400 for 14-in. flywheel ring gear.	1

Ignition Coil

PART NO.	DESCRIPTION	QTY.
1115207	Coil, transistorized-ignition. (Embossed 176-12V) (1.8-ohm wire). Part of ignition kit 3997782.	1
10037380	Coil, HD electronic-distributor, oil-filled. Use w/distributor 10037373 and control box 10037378. Coil has alkyd top w/tall high-tension tower to prevent arcing.	1

Sparkplug Wire

PART NO.	DESCRIPTION	QTY.
8914472	Sparkplug-wire kit, straight-boot type, solid-core, high-silicon cover. For all except HEI distributor.	1

Ignition Distributor, Components

PART NO.	DESCRIPTION	QTY.
1111263	Distributor Assembly, transistorized w/ball-type shaft. For gear-drive cam. Use on chain-drive cam by changing lower gear to 1958599. Use w/ignition kit 3997782.	1
1111267	Distributor Assembly, transistorized w/bronze bushing-type shaft. Use w/ignition kit 3997782.	1
1110985	Distributor Assembly, dual-breaker point w/tach drive.	1
1103302	Distributor, Cap-and-Coil Assembly, HEI.	1
14044871	Distributor Assembly, electronic, non-tach-drive-type, w/tuftrided shaft, sintered bushings, hardened and concentric-ground advance cam, high-output magnetic pickup, locked-out vacuum advance and reinforced weight pins. Uses std. cap and rotor.	1
1941551	Cap, distributor, screw-type. For dual-point distributor.	1
1971244	Cap, distributor. For electronic distributor 14044871.	1
1960779	Pole Piece, rotating, w/pick-up coil. (Can be modified to repair distributor 1111263 or 1111267).	1
1964272	Pole Piece, stationary, w/weight base. (Can be modified to repair distributor 1111263 or 1111267).	1
3865886	Shaft, oil-pump-to-distributor.	1
3997782	Transistor Ignition Unit. Includes 3955511 amplifier, 1111507 coil, 6288704 connector, 6297793 or 8901973 wire, and 6297688 harness. Use w/distributor 1111263 or 1111267.	1
3955511	Amplifier, pulse, transistor-ignition. Part of Unit 3997782.	1
10037378	Amplifier, HD electronic-ignition. Use w/distributor 14044871 and coil 10037380.	1
8901973	Wire, transistor ignition.	1
6297688	Harness, transistor-ignition pulse amplifier. Part of unit 3997782.	1
10037377	Harness, HD electronic-ignition, amplifier-to-distributor, dash-mounted unit. Use w/distributor 1404871 and coil 10037380.	1
10039932	Harness, HD electronic-ignition, amplifier-to-distributor, engine-compartment-mounted unit. Use w/distributor 1404871 and coil 10037380.	1
10037376	Changeover Switch, HD electronic-ignition, dash-mounted. Provides for operation of back-up ignition. Requires two 10037378 amplifiers. Use w/distributor 14044871.	1
10037379	Rev Limiter, soft-touch, HD electronic-ignition. Plugs into ignition module. Use w/distributor 14044871.	1
10039933	Module Kit, rev limiter, 5000 series. Use w/limiter 10037379. Includes five modules: 5000, 5200, 5400, 5600 and 5800 rpm.	1
10039934	Module Kit, rev limiter, 6000 series. Use w/limiter 10037379. Includes five modules: 6000, 6200, 6400, 6600 and 6800 rpm.	1
10039935	Module Kit, rev limiter, 7000 series. Use w/limiter 100373379. Includes five modules: 7000, 7200, 7400, 7600 and 7800 rpm.	1
10039936	Module Kit, rev limiter, 8000 series. Use w/limiter 100373379. Includes five modules: 8000, 8200, 8400, 8600 and 8800 rpm.	1

Manifolds

PART NO.	DESCRIPTION	QTY.
3933163	Intake Manifold, closed plenum for Holley carburetor.	1
3931093	Shield, oil, for bottom of intake manifold.	1
3955528	Gasket Kit, manifold-to-cylinder head, all 396/427/454 CID.	1

Air Cleaner

PART NO.	DESCRIPTION	QTY.
6423907	Air Cleaner, open-element, w/14-in. chrome cover. Includes base 6422188, cover 6421832 and element 6421746 or (A21ZCW).	1

Fuel Pump

PART NO.	DESCRIPTION	QTY.
6415748	Fuel Pump, high-capacity.	1
3704817	Pushrod, fuel-pump.	1

Tuning hints
with a little extra attention your engine will really sing

And you think you have problems tuning one car? How'd you like to keep four and sometimes five race cars in competitive tune for the no-holds-barred AHRA drag-racing circuit? Bill Hielscher from Irving, Texas does just that and takes home his share of wins. Job takes 26 hours a day.

THE TUNEUP

Just because you build a 650-horsepower big-block Chevy is no insurance that you'll have within 100 horsepower of that reading one month after the engine goes to work in the car. Because of the enormous horsepower available from the large-block Chevy and the fact horsepower goes away gradually in normal driving, it is difficult to tell that your engine is not performing at its peak. After a sharp tuneup, you'll realize the engine had lost much of the potential response which was built into the engine.

Most modified engines never produce the horsepower of which they are capable because their builders fail to spend the extra effort required to make their creations run perfectly. You can make your investment pay maximum dividends through careful attention to details during assembly—and then tuning it to perfection after you have installed it—by using a chassis or an engine dyno. A few hours of chassis or engine-dyno time will produce far more performance than you might imagine.

Items which deserve special attention include carburetor jetting, linkage, and synchronization (for dual carbs). Ignition timing—initial setting and centrifugal-advance rate and amount—is another area of importance. Transmission and rear-axle ratios must also be carefully selected to get the desired performance.

Multiple-carb installations do not stay synchronized, so buy a Uni-Syn or other synchronizing tool—and use it frequently, or switch to a single four-barrel with the correct air-flow capacity for your engine. Prior to any tuning or competition efforts, check that the carb linkage works easily to open the carb/s fully and close them positively with no binding or sticking. This kind of action is well worth whatever time it takes to make it happen.

No gaskets should obstruct the carbs at their bases or the manifold castings where they join the heads.

Selection of the correct jets can be accomplished by observing the color of the porcelain insulators. However, new plugs may take time to "color"—so use an illuminated magnifier (Champion or A-C tool) to make this chore an easier one. These illuminate the base of the porcelain where it is buried in the plug shell. Because this is the first part of the plug to "color," the magnifier should be considered an essential tool for any racer's toolbox.

When you have installed the jets which give maximum performance (12:1 air:fuel ratio), you may have less economy than you'd like for daily driving. Should this be the case, use leaner jets for around-town driving and highway use—remembering that the engine is being fed a too-lean mixture for full-throttle operation. Re-jet the carbs when it's time to race. Part-throttle and cruise air/fuel ratios are usually set for 14 or 15:1—or even leaner on many cars which are set up for minimum emissions.

Air cleaners can affect mixture, so run your tests with the air cleaner/s installed. Dyno tests with and without the cleaners can quickly show whether the air cleaner/s are restrictive. The air cleaner base must always be left installed to provide correct air entry to the carburetor.

Set the ignition timing per factory specs, using added initial advance with caution as described in the chapter on ignition. Avoid setting the timing while the engine is being tested on a chassis or engine dyno—sometimes called "power timing" *unless you load the engine long enough for the temperatures to stabilize.* Flash readings made in conjunction with quick twisting of the distributor to get impressive readings are not the answer. Should you do this, and then run the engine with the same setting—especially at high speeds on the road or in competition—you can destroy the engine. Here's what happens—the spark setting which can be tolerated by the engine during a flash reading before spark-plug and cylinder-head temperatures have stabilized is more advanced than the engine can safely use after temperatures have reached the higher values in steady operation. As a result, the *dyno setting of ignition timing for the highest flash reading will always be too far advanced.* Using such settings for anything more than a quick blast through the quarter-mile will cause destructive detonation.

Both good and bad news are foretold in this picture. A common practice to warm up the lubricant in the entire running gear is to jack the car up at the differential housing, open the hood and set the engine for 2000 RPM idle. This is fine... unless someone should lean against the car and either rear wheel touches the ground. Better place stands under two points at the rear instead of just one if you'd rather be safe than sorry.

Notice that: A. There is no fan shroud. B. The radiator is "half-thickness." C. A Flexolite fan is used. D. That an outside mass Dayco fan belt is used. All of this is part of the drag-racing game.

Tuning pays off in performance dividends. This is true with any engine, whether it is stock or highly modified. You may not have access to a dyno. Even so, you can do a lot of tuning on the road without traffic and with easy-to-see markers. Keep detailed records in a notebook—and *never change more than one thing at a time.* You'll be surprised at how accurately you can spot what is helping—or hindering—your car's performance. You can equalize air-density effects by testing during the coolest part of the day—and preferably on days with similar barometric pressures.

TOOLS REQUIRED

To perform a good tune-up you need a minimum of special tools—and a maximum of common sense. Let's start with the tools—you'll have to take care of the other. In order of importance we'd have to vote for a compression gage, a vacuum gage and a tachometer. Feeler gage, a spark-plug wrench, and various small wrenches and screw drivers should round out the list of items necessary to take care of a Saturday-morning tuneup.

HOW TO USE COMPRESSION GAGE

Let's begin with the compression gage—a quick and accurate way to discover if one cylinder is "laying down" on you. A compression gage can be used to discover broken valve springs, a blown head gasket, a crack in the cylinder wall, a broken ring or one that has lost its spring and bent or burned valves. In other words you either start by running a compression check on the engine—or spend a month of Sundays changing plugs, wires, carburetors, coils, points, beating on the fender and swearing Chevys are no good. A long parts bill later you'll discover that a broken valve spring accounts for the sour performance.

Use compressed air (a bicycle pump will do) to blow away any trash or debris that has collected around the base of the spark plugs. Disconnect the wires and remove all of the spark plugs. Pull the coil-wire out of the distributor and jam the carburetor linkage to full-open position. The way to go here is to remove the throttle return spring and then block open the linkage. Often this can be done by just wedging a screwdriver into some part of the linkage to prevent its movement. Number a piece of paper 1 through 8. Insert the compression-gage fitting into a spark-plug hole (#1). Have wife, buddy, or neighbor crank the engine over at least five revolutions. Record the gage reading. Release the pressure in the gage and repeat the procedure to read the same cylinder again. Record the reading beside the original reading. Move on to the next cylinder and repeat the double-reading procedure at each cylinder.

Watch the gage carefully as the engine is cranked. The needle should rise steadily as the piston rises in the cylinder. If the needle does not rise steadily on one or more cylinders suspect a stuck or sticking valve... either intake or exhaust. If the readings are about equal between cylinders but all are lower than the specifications call for the problem is most likely badly worn rings or leakage around the valve seats. To narrow this down, squirt a small amount of engine oil into the spark plug hole of one of the cylinders to help the rings seal. Then take another compression reading on this cylinder. If the reading shows a substantial rise over the previous readings then the compression leakage is past the piston rings. If the oil has no effect on the reading, then the leakage is past the valves. They could be bent, burned or a valve guide could be severely worn. If one cylinder reads much lower than the rest, check for a broken valve spring if the cylinder fails to respond to the oil squirt. If adjacent cylinders show a very low reading compared to readings obtained from the other six then you've probably got a blown head gasket on your hands.

If compression readings are above those specified for the engine the combustion chambers are most likely covered with a thick coat of carbon from bad fuel, low-speed operation, general neglect or overly rich jetting. If you suspect a blown head gasket, put everything back in working order, fire the engine and pull the radiator cap. Often the ailing gasket will allow a stream of air bubbles to flow into the cooling system and these are readily seen when the engine is fired. Be quite careful in pulling a head off, if you do suspect a blown head gasket. Try to save the gasket—NOT FOR REUSE—but for examination. The trouble might not be a blown gasket, but a cracked cylinder wall or combustion chamber.

HOW TO USE A VACUUM GAGE

A vacuum gage is an inexpensive tuneup tool that can be used to locate trouble while you are doing daily driving. It also doubles as a tuneup tool. If you've never used a vacuum gage on an engine before, hang on to the instructions that come with it because they'll usually include a chart and you'll be able to get much more from the gage if you'll commit some of the needle movements to memory. One profitable way to spend an afternoon is to connect a vacuum gage to an engine which you know to be in good tune and then make measured changes to the valve lash, or ignition timing or idle mixture to see what the gage does under each "bad" condition. Keep track of these changes on a piece of paper so that you can return the engine to its original state of tune—with or without the aid of the vacuum gage. Always allow the engine to reach normal operating temperature before using the vacuum gage.

HOW TO USE YOUR TACHOMETER IN TUNEUPS

Let's say you're in a situation where the engine suddenly starts running rough or there is a noticeable loss of power. A compression gage is not handy, but you have a tachometer. Short each spark plug wire to ground one at a time and note how much RPM drops on the tach . . . or pull off the plug wires one at a time. A leak in a cylinder caused by rings, valve seats or cracks in the cylinder wall or combustion chamber will allow pressure to drop within the cylinder. That cylinder produces less power, affecting RPM.

A cylinder doing its share of work without serious leakage will drop 50 to 100 RPM when that plug wire is pulled off or grounded. If the tach shows a decrease in RPM of only 10 to 20 RPM, you've found a sick cylinder.

AIR LEAKS

Rough idle is usually blamed on bad plugs or plug cables. But, once in a while, the rough idle persists after the obvious things are corrected. In such cases,

Bill Jenkins goes after a spark plug on his Pro Stock car. Manifold is Edelbrock TR-4X model. Foam-rubber air seal is typical of Jenkins' thoroughness. Keeps engine-heated air away from carbs so they ingest only cool air from hood vent.

according to Champion engineers, you should always suspect air leakage into the intake manifold. The first step is to check the torque on the main manifold bolts and the carburetor base. But, don't overlook vacuum-hose connections such as heater/defroster controls, spark-advance connection to the distributor and other vacuum connections which are used for various control functions. Sometimes the hose itself will leak—and other leaks occur around stripped threads of manifold or carburetor fittings. A hose sometimes cracks on its underside where it's not immediately obvious.

STICKY LIFTERS

A sticking valve or lifter will often be the result of "getting into the throttle" on a Saturday night after several weeks of driving the car slowly on city streets. Fill a squirt can with carburetor cleaner and squirt small amounts of the liquid down the push rod toward the stuck lifter or around the valve stem while the engine is idling. Any number of carburetor cleaners which can be used while the engine is running at a fast idle can be effective remedies for unsticking a valve . . . but pouring such stuff in the carburetor won't do anything except create smog to annoy your neighbors.

Weaver Brothers' drive uses a cogged belt to operate the water pump when only the pump is used. Other drives provide for fuel-injection pumps and the Weaver dry-sump assemblies.

ENGINE TEMPERATURE vs. PERFORMANCE

A cold engine sometimes feels stronger because the underhood temperatures are down and the carburetor is getting cooler air. If the engine is assembled correctly, it should run as well hot or cold, with less wear if it is warm. Chevrolet claims no difference in power between 145 and 195 degrees and they never run cold engines on the dyno.

THERMOSTATS

Don't let any would-be automotive expert talk you into taking out the thermostat. Thermostatically controlled cooling is an essential feature and only the stupid person who doesn't understand what is happening takes it out. If a part can be left out of an engine safely, you can be sure that the manufacturer won't put it on in the first place. Auto engineers always go by the slogan of the famous GM engineer, Boss Kettering, "Parts left out cost nothing and cause no service problems." If there was any way for Chevy's engineers to ensure that the engine got fast warm-ups to avoid sludge and acid formation—and therefore engine wear—other than by controlling the water temperature, you can be sure that they would do it in a minute. And, removing the thermostat can reduce the cooling which you get from the water circulation. It's also possible that taking out the thermostat will reduce performance because more HP is required to drive the water pump when the restriction is taken out.

Bear in mind that engine wear is increased when warm-up time is increased. A quickly warmed-up engine suffers less wear, especially when it is allowed to reach and maintain its normal operating temperature for a period of time prior to shutting it off. Short hops during which an engine never warms up increase oil dilution by gasoline, build up varnish and sludge accumulations, and greatly increase cylinder wear. High wear rate is caused by combustion products which condense on the cylinder walls to cause etching and rapid wear.

Cold cylinders also mean thick oil which increases the friction that pistons must overcome. The hotter the cylinder walls, the less friction loss.

You can check this oft-proved and well-documented series of facts in any internal-combustion-engine textbook.

Cooling System

The cooling system should also be checked when you are going about your tune-up chores. Before starting the engine—while everything is still cold—check the coolant level. Models without a separate supply (header) tank should be maintained with the coolant level just below the top of the filler neck when cold. If a separate supply (header) tank is in the system—as on Corvettes—the level should be maintained at the half-full mark in the tank.

If you are filling the system for the first time, you may notice an overflow the first time that you get the engine up to operating temperature. This is perfectly normal and indicates that the system is leveling out due to expansion of the coolant. The level should remain where it belongs from then on . . . slightly below full in any case.

Engine overheating can occur even with a well-maintained cooling system if some other factors are overlooked. Radiator cleanliness is important. Bugs collected in the radiator core during the summer reduce the air flow enough to cause overheating. A high-pressure air hose directed at the back of the radiator will usually clean the little rascals out. It is a good idea to do this every fall.

Because the oil picks up and radiates a lot of engine heat, a low oil level can also cause a higher water temperature. A dirty engine and a thickly coated oil pan can hold a lot of heat and add to the cooling problem. And, an incorrectly adjusted ignition or Transmission Controlled Spark System can also cause overheating.

The structure around the radiator is extremely important to the radiator's overall cooling capabilities. Shrouds at the back side of the radiator allow the fan to work more efficiently. Directional panels at the front of the radiator funnel the air through the radiator, using the forward motion of the car to assist in ramming air through the radiator.

These must be intact and correctly installed with the appropriate seals.

Thermostats. If they stick closed (cold-engine position) the engine will run too hot. If they stick open (hot-engine position) the engine will run too cold in anything except hot weather. All big-block engines are equipped with 170° thermostats except 1966-69 which had 195° thermostats if the car was equipped with air conditioning. In 1970, almost all of the big blocks had 180° thermostats. 1971 454's had 180° thermostats and the rest were equipped with $195^\circ F$ units.

IGNITION CHECKS

With these guidelines on how to locate hidden trouble you've got a sound foundation for a good tuneup. Follow through by cleaning, gapping and replacing the plugs, or replacing them with new ones. Always check the gap on new plugs before installing them. While you're shoving the protective boots over the plugs, check the rubber or Neoprene for cracks or insulation damage which could allow voltage to leak off. Do the same thing with all of the primary wiring. Then as a double check on wiring, run the engine at night with the hood open to locate any arcing from wire to wire, or wire to ground.

Install new points and condenser or inspect the points for pitting and adjust them for the correct gap. Take a close look at the distributor cap while it's off. Are the terminal posts corroded, burned or loose? These ailments and others related to the distributor cap can lead to some difficult-to-locate problems.

SPARK PLUG READING

Learning to read spark plugs is fundamental to keeping track of the performance of any engine. The guidelines on spark-plug reading in the ignition chapter should get you started. Spend plenty of time developing this skill of "reading" and if you believe it too difficult to bother with, keep in mind that this one skill separates the men from the boys at Indianapolis where alcohol and nitromethane fuel make plug reading a revered science!

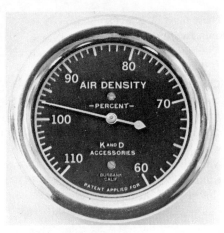

K-D air density meter is an essential tool for the serious tuner.

OTHER ITEMS TO CHECK

Don't overlook the ordinary in a tuneup. Now is the time to inspect the fan and various accessory belts for wear and cracking. Starting at the fuel-pump fittings, visually move along the fuel line to the carburetor, checking the fittings for gas stains that would indicate leaks. If flex hose is used, check it very carefully for cracks or evidence of seepage where the flex line joins the steel fitting. Inspect the radiator hoses—especially the lower one because it is often overlooked until it's too late to change at your convenience.

Remove corrosion buildup from the battery, terminals and starter connections with a mixture of household baking soda and warm water.

If you're running a set of headers on the car, check each bolt to see that it is tight. On a new set of headers this should be done at least once a week for several weeks until the header flange takes a "set." Then the bolts will remain tight for longer periods of time. At this time, they can be removed one at a time, touched with a couple of drops of Loctite and tightened down.

Pull the valve covers, remove the breathers and clean both with solvent. They can soak while all of the head bolts are checked for the correct torque loading and the valve lash is adjusted. When installing new gaskets, apply your favorite gasket cement to one side (the side that goes against the valve cover) and stiff wheel bearing grease along the other side of the gasket. The valve cover and gasket may now be removed as a unit several times before the gasket need be replaced.

DZUS of West Islip, Long Island, New York now makes a special fastener for fast removal and positive installation of covers. One-quarter turn with a screwdriver releases each fastener. No machine work is required for installation.

AIR DENSITY GAGE

One of the most overlooked but necessary tools for tuning for drags or road racing is the air-density meter. Air density changes from hour to hour — day to day — and most certainly from weekend to weekend. With the info provided by the makers of this device, you can change jets according to the meter. No longer do you have to arrive at the right jets by trial and error. You get it right the first time. If you're gonna race, get one!

REV LIMITS

As you go about your tuning efforts, stick to Chevy's recommended RPM limits for their high-performance big blocks. Maximum speeds which they have indicated are different for the 396/402/427 than for the 454 because the longer stroke produces critical piston speeds at lower RPM's.

Engine	Track Racing	Drag Racing
396/402/427	6600-7000 RPM	7600 RPM
454	6200-6600 RPM	7200 RPM

"FREE HP"

Although variable-pitch fiberglass fans are available for the big-block engines, the Chevrolet clutch fan with a thermo-viscous hub is the best deal because the engine never couples to it unless the thermostat control says that things are getting too warm. Here again, the factory item is the desirable, hot setup, good-guy part to use. The popular fiberglass variable-pitch fans are extremely noisy, so try to listen before you lay out your cash for one of them.

For drag racing you can eliminate the fan or take off the belt, at least in some classes. If a working alternator is not required by the rules, remove the belt for the race and simplify the entire procedure . . . and get even more horsepower.

HOW TO COMPENSATE FOR THE AUTOMATIC-RETARD CHARACTERISTICS OF CHEVROLET's TRANSISTOR IGNITION

When you use the Chevrolet mag-pulse-triggered ignition system, there is an effect which you must know about if you want to maintain full top-RPM performance. Constant switching time delays in the transistor system create an automatic retard of $1°$ per 1000 RPM (engine) *after full mechanical advance has been reached.* You may have noted the effect with your timing light. The advance appears to be "backing down" as you rev the engine. This is exactly what happens and it will reduce top-end HP.

To end up with the full advance you want at peak RPM, run the engine to the point where all mechanical advance is in. If your timing light shows this as $38°$ at 4000 RPM engine speed (watching your degreed harmonic balancer and tach), and you want $38°$ at 7000 RPM, you'll have to crank in an additional $3°$ BTDC on the initial setting. Now your timing light should show $41°$ advance at 4000 RPM. The ignition system will automatically retard $3°$ to give the desired $38°$ at peak RPM. Chances are good that the over-advance in the mid-range will be helpful during acceleration.

If you have set up the distributor so that all advance is in by 3000 engine RPM — as is often done for drag racing — check with the timing light to see where the advance is all in. It could be somewhat higher/lower than you thought, so do this with the engine running. If you have $38°$ at 3000 and expect to turn 7000 "in the eyes," the spark at that RPM will be $4°$ retarded.

You would have only $34°$ advance instead of the $38°$ you wanted. To compensate, increase the initial advance setting by $4°$ and use your timing light to check that you have $42°$ at 3000 RPM. The automatic-retard characteristic will back it off to the desired $38°$ at 7000 as we have indicated.

This same effect may occur with other mag-pulse or point-triggered transistor ignition systems. Be sure to check for this with your timing light. This sort of behavior may not be observable on distributor machines which do not test the entire system.

More rev limiters and tachometers: page 124

Big-block suppliers

Accel
Box 142
Branford, CT. 06405

Aviaid Metal Products
7570 Woodman Pl.
Van Nuys, CA 91405
818/786-4025

Baker Engineered Racing Engines
14122 Ironwood NW
Grand Rapids, MI 49504
616/677-5234

Gale Banks Engineering
546 Duggan
Azusa, CA 91702
818/969-9605

Booth-Arons Inc.
3861 W. 12 Mile Road
Berkley, MI 48072
313/398-2730

Brodix Inc.
P.O. Box D
Mena, AR 71953
501/394-1075

Brooks Racing Components
7091 Belgrave Ave.
Garden Grove, CA 92641
714/893-0595

Cam Dynamics
4031 Winchester Rd.
Memphis, TN 38118
901/794-2870

Carrillo Industries
33041 Calle Perfecto
San Juan Capistrano, CA 92675
714/493-1230

Childs & Albert
11030 Sherman Way
Sun Valley, CA 91352
818/765-0988

Cloyes Gear & Products, Inc.
4520 Beidler
Willoughby, OH 44094
216/531-3264

Competition Cams
2806 Hangar Rd.
Memphis, TN 38118
901/795-2400

Competition Fuel Systems
3820 E. 44th, Unit 410
Tucson, AZ 85713
602/745-1691

Competition Grinding
2712 N. Columbus
Tucson, AZ 85712
602/323-1578

Crane Cams, Inc.
100 NW 9th Terrace
Hallandale, FL 33009
305/457-8888

Crower Cams
3333 Main St.
Chula Vista, CA 92011
619/422-1191

Diamond Racing Products
23003 Diamond Drive
Mt. Clemens, MI 48043
313/792-6620

Earl's Supply
825 E. Sepulveda
Carson, CA 90745
213/830-1620

Edelbrock Equipment Co.
411 Coral Circle
El Segundo, CA 90245
213/322-7310

Enderle Fuel Injection
1282 S. Los Angeles Street
Glendale, CA
213/243-2175

Sig Erson Cams
550 Mallory Way
Carson City, NV
702/882-1622

Fischer Engineering
9003 Norris Ave.
Sun Valley, CA 91352
818/504-0030

George Foltz Racing
16521 Eastland
Roseville, MI 48066
313/779-9040

Gaerte Engines
615 Monroe St.
Rochester, IN 46975
219/223-3016

General Kinetics Co.
5161 Trumbull
Detroit, MI 48208
313/832-7360

Ed Hamburger's Hi-Performance
1590 Church
Toms River, NJ 08753
201/240-3888

Hank the Crank, Inc.
7253 Lankershim Blvd.
North Hollywood, CA 91605
818/765-3444

Hedman Headers
9599 W. Jefferson Blvd.
Culver City, CA 90230
213/839-7581

Helm, Inc.
Publications Division
P.O. Box 07130
Detroit, MI 48207

Hilborn Fuel Injection
25891 Crown Valley Parkway
South Laguna, CA 92677
714/831-1170

Holley Carburetors
11955 E. Nine Mile Rd.
Warren, MI 48090
313/497-4000

Hooker Headers
1032A W. Brooks St.
Ontario, CA 91762
714/983-5871

Honest Charley, Inc.
108 Honest St.
Box 8429
Chattanooga, TN 37411
615/892-2114

Iskenderian Cams
16020 S. Broadway
Gardena, CA 90248
213/770-0930

Jenkins Competition
153 Penn Ave.
Malvern Industrial Park
Malvern, PA 19355
215/644-4777

Jet Engineering, Inc.
5300 Aurelius Rd.
P.O. Box 25066
Lansing, MI 48909
517/393-5110

Joehnck Racing Heads
360 S. Fairview, Unit C
Goleta, CA 93117
805/964-7759

Katech Inc.
22969 Rasch
Mt. Clemens, MI 48043
313/791-4120

Kinsler Fuel Injection
1834 Thunderbird
Troy, MI 48084
313/362-1145

Klein Engines
4020 E. Elwood
Phoenix, AZ 85040
800/237-8849

Lakewood Industries
8700 Brookpark Rd.
Brooklyn, OH 44129
216/398-8300

Lunati Racing Cams & Equipment
3871 Watman Avenue
Memphis, TN 38118
901/365-0950

Manley Performance Engineering
13 Race St.
Bloomfield, NJ 07003
201/743-6577

McLaren Engines, Inc.
32233 W. Eight Mile Rd.
Livonia, MI 48152
313/477-6240

Milodon Engineering Co.
9152 Independence Ave.
Chatsworth, CA 91311
818/882-6422

Moldex Crankshaft Co.
25249 W. Warren Ave.
Dearborn Heights, MI 48127
313/561-7676

Moroso Performance Products
80 Carter Drive
Guilford, CT 06437
203/453-6571

Motion Performance Inc.
598 Sunrise Hiway
Baldwin, NY 11510
516/223-3172

Norris Cams
14762 Calvert Street
Van Nuys, CA 91401
818/780-1102

Peterson Automotive Research
4485 S. Broadway
Englewood, CO 80110
303/781-7290

Ed Pink Racing Engines
14612 Raymer
Van Nuys, CA 91405
818/873-3460

Prototype Engineering
140-B Shepard St.
Wheeling, IL 60090
312/459-0611

Racer Brown Cams
9270 Borden Ave.
Sun Valley, CA 91352
818/897-4044

Racing Head Service
2806 Hangar Road
Memphis, TN 38118
901/794-2830

Reed Cams
Dekalb Peachtree Airport
Building 35B
Atlanta, GA 30341
404/451-5086

Rodeck Inc.
18093 S. Figueroa
Gardena, CA 90248
213/583-5791

Ryan Falconer
1370-B Burton Ave.
Salinas, CA 93901
408/758-8434

Shaver Specialities
6127 S. Western Ave.
Los Angeles, CA 90047
213/752-3713

Sissel Racing Heads
2116 Seaman Ave.
S. El Monte, CA 91733
818/443-5028

Speed Pro (Sealed-Power)
100 Terrace Plaza
Muskegon, MI 49443
616/724-5411

Sperry Fuel Controls
15909 E. 14 Mile Rd.
Fraser, MI 48206
313/293-7375

Stahl Headers
1515 Mt. Rose Ave.
York, PA 17403
717/846-1632

Summers Brothers
530 S. Mountain Avenue
Ontario, CA 91761
714/986-2041

SuperFlow
3512-E North Tejon
Colorado Springs, CO 80907
303/471-1746

Traco
11928 W. Jefferson Blvd.
Culver City, CA 90230
213/398-3722

TRW, Inc.
8001 Pleasant Valley Rd.
Cuyahoga, OH 44131
216/447-8164

Valley Head Service
19340 Londelius St.
Northridge, CA 91324
818/993-7000

Weaver Brothers, Ltd.
1980 Boeing Way
Carson City, NV 89701
702/883-7677

Weiand
2737 San Fernando Rd.
Los Angeles, CA 90065
213/225-4138

Acknowledgements

Lance "Buzz" Morris of Santa Ana, Calif., was one of the major contributors of construction information for our book. He's shown here twisting the crank and contemplating how to get still more HP out of one of his very successful big-block racing engines.

"Building a book" is a difficult and exacting task thoroughly laced with all of the time-consuming and aggravating problems that plague those who try to build "right-on" engines. The job of putting this book together was eased considerably by a number of men who were very interested in seeing—at last—a "right-on" book about the big-block Chevy.

Our deepest dept of gratitude is owed to the people at Chevrolet. Walter MacKenzie, Technical Projects Manager for Chevrolet Public Relations, helped with specifications, photos and details that we could never have obtained anywhere else. And, GM photographer Ed Sperko's beautiful front-cover photograph should gladden the hearts of every big-block enthusiast. Chevrolet's Product Promotion Group answered countless questions. Bill Howell, Engineer and Big-Block Specialist in that organization, deserves special thanks for his advice and for steering us away from expensive and trick ideas and procedures which were not necessary.

Lance (Buzz) Morris spent countless hours cluing us in on the "do's and don'ts" of big-block building. Morris was instrumental in rounding up components for pictures and in reading and correcting our writing attempts. Bob Joehnck of Santa Barbara helped in the distillation process and kept after us with "what the guys really want to know is . . ." Many tips from crack engine builders Dave Diamond, Ron Hutter, Jere Stahl, Jim Cavallero, Paul Hogge, Bill King and Jim Kinsler are in this book. All have tried countless big-block combinations and were free with information. Not only that, they were helpful and patient in answering our questions and always willing to "hold still" for just one more picture. Kay Sissell and his flow bench provided a bunch of invaluable information on how to get more air in—and thus more horsepower out of—big-block heads. Racer Brown provided much of the camshaft/valve-train information. Doug Roe aided with carburetion details.

High-Performance Specialist Dino Fry and Parts Manager Don Zanstra of Courtesy Chevrolet in San Jose, California are a couple of those unsung heroes who make building a hot engine—and a hot book—possible. They cross-checked and tracked down elusive high-performance parts numbers, some of which seemed to change almost hourly. And their extensively complete stock of high-performance parts allowed us to show parts which have never been previously pictured.

Ex-Team-McLaren engine man George Bolthoff helped with solid information on what parts he'd found to be reliable in Can Am cars and exploded a lot of the false ideas we'd held about what was required to make these engines live at high RPM's for hundreds of miles at a time.

Freelancers such as Wayne Thoms came to our rescue time and again with needed photos, as did several magazine editors, notably Lee Kelly at Popular Hot Rodding.

Others who helped with specific advice and information included Don Gonyou and Andy Guria of Holley Carburetors, Jim McFarland of Edelbrock Equipment and Jerry Thompson of Troy Promotions. We have undoubtedly left out names of others who helped—but we didn't mean to. Everyone with whom we talked was enthusiastic and helpful with the project of putting a book into print which would "tell it like it is" about a formidable and highly respected piece of equipment—the big-block Chevrolet.

As you peruse these pages, you'll notice an unusual absence of speed equipment. This is intentional. Our feeling is that with so much very good Chevrolet equipment available—why bother with unknown and unproved qualities and quantities and end up with countless combinations which won't run as strong as a stock-out-of-the-crate engine? Don't misunderstand, please! We are not saying that all speed equipment is "BAD" —just that quite a bit of it is completely useless on street-driven cars. Our intention in writing this book was to provide a base from which to build a strong, reliable powerplant—not a base from which to spend money needlessly. We started with this premise and the more knowledgeable engine builders we talked with, the more we knew that we were on the right track.